What Is Biodiversity?

DATE DUE

What Is Biodiversity?

James Maclaurin and Kim Sterelny

The University of Chicago Press   CHICAGO AND LONDON

JAMES MACLAURIN is a senior lecturer in the Department of Philosophy at the University of Otago, New Zealand, and has also been a Marsden Post Doctoral Fellow at Victoria University. He is the author of numerous articles published in professional journals.

KIM STERELNY divides his time between Victoria University of Wellington, where he is a professor of philosophy, and the Research School of Social Sciences and the Centre for Macroevolution and Macroecology at the Australian National University. He is the editor of the journal *Biology and Philosophy*, and his books include *Evolution of Agency and Other Essays; Thought in a Hostile World; Dawkins vs. Gould*; and, with Paul Griffiths, *Sex and Death: An Introduction to Philosophy of Biology*, which is published by the University of Chicago Press.

The University of Chicago Press, Chicago 60637
The University of Chicago Press, Ltd., London
© 2008 by The University of Chicago
All rights reserved. Published 2008
Printed in the United States of America

17 16 15 14 13 12 11 10 09 08   1 2 3 4 5

ISBN-13: 978-0-226-50080-5 (cloth)
ISBN-13: 978-0-226-50081-2 (paper)
ISBN-10: 0-226-50080-2 (cloth)
ISBN-10: 0-226-50081-0 (paper)

Library of Congress Cataloging-in-Publication Data
Maclaurin, James.
    What is biodiversity? / James Maclaurin and Kim Sterelny.
        p.   cm.
    Includes bibliographical references and index.
    ISBN-13: 978-0-226-50080-5 (hardcover : alk. paper)
    ISBN 10: 0-226-50080-2 (hardcover : alk. paper)
    ISBN-13: 978-0-226-50081-2 (pbk. : alk. paper)
    ISBN-10: 0-226-50081-0 (pbk. : alk. paper)
    1. Biodiversity. I. Sterelny, Kim. II. Title.
QH541.15.B56M325   2008
333.95–dc22

⊗ The paper used in this publication meets the minimum requirements of the American National Standard for Information Sciences—Permanence of Paper for Printed Library Materials, ANSI Z 39.48–1992.

For Kristen and George and Melanie and Kate

Contents

Acknowledgments

This book is the product of a Marsden Fund grant. The full-time focus the grant made possible kick-started the project. Even so, it has been a long project, and we have received help from many quarters.

Our special thanks go to Ben Jeffares and Russell Brown for dedicated research assistance during the early phase of the project. James Maclaurin would like to thank Alan Musgrave for being extraordinarily helpful and supportive and also the Department of Philosophy at the University of Otago, which continues to be a stimulating and enjoyable place to think. Kim Sterelny would like to thank his two academic homes, the philosophy programs at Victoria University of Wellington and the RSSS, ANU. Both continue to be collegial and supportive environments in which to do empirically oriented philosophy.

Many people were happy to be used as sounding boards and provided useful feedback. These include Nick Agar, Peter Anstey, Jochen Brocks, David Braddon-Mitchell, Lindell Bromham, Brett Calcott, David Chalmers, Geoff Chambers, James Chase, Colin Cheyne, Mark Colyvan, Tim Dare, Kath Dickinson, Steve Downes, Heather Dyke, Patrick Forber, Peter Godfrey Smith, Todd Grantham, Paul Griffiths, Mike Hannah, Frank Jackson, Ben Jeffares, Richard Joyce, John Matthewson, George McGhee, Andrew Moore, Karen Neander, Daniel Nolan, Samir Okasha, Charles Pigden, Josh Parsons, Gerhard Schlosser, Nick Shea, and Daniel Stoljar.

Secretarial support at Otago has been ably provided by Sally Holloway and Kate Anscombe. The index was prepared by Meg Cordes.

We reserve special thanks for Christie Henry at the University of Chicago Press, who has been unflaggingly helpful, and for our excellent copy editor, Dawn Hall.

Kim thanks his partner Melanie for her unfailing support of his research efforts despite her own intense research commitments, and his daughter Kate for making life much easier through her cheerfulness and goodwill. James would like to thank his wife Kristen and son George for their love and inspiration, and he extends grateful thanks to his coauthor, a mentor and friend without whom this book would not have happened.

1 *Taxonomy Red in Tooth and Claw*

Eliminate one species, and another increases in number to take its place. Eliminate a great many species, and the local ecosystem starts to decay visibly. Productivity drops as the channels of the nutrient cycles are clogged. More of the biomass is sequestered in the form of dead vegetation and slowly metabolising, oxygen-starved mud, or is simply washed away. Less competent pollinators take over as the best-adapted bees, moths, birds, bats, and other specialists drop out. Fewer seeds fall, fewer seedlings sprout. Herbivores decline, and their predators die away in close concert.   *Wilson 1992, 14*

"Biological diversity" means the variability among living organisms from all sources including, inter alia, terrestrial, marine and other aquatic ecosystems and the ecological complexes of which they are part; this includes diversity within species, between species and of ecosystems.

> *UN Conference on Environment and Development, Rio de Janeiro, 1992, Convention on Biological Diversity, Article 2*

1.1 BIODIVERSITY AND "BIODIVERSITY"

Concepts of diversity, we shall argue, are important in many areas of biology. But "biodiversity," the term, comes to us from conservation biology. In 1992, Edward O. Wilson wrote *The Diversity of Life*. The aim of the book was to draw attention to species loss and in particular to the loss of species caused by human activities. It was not a new message. Wilson was one of a group of prominent ecologists who had been warning of a massive human-caused extinction event since the 1970s. The prognosis was dire. In *The Sinking Ark*, Norman Meyers (1979, 4–5) suggested that we might be losing as many as 40,000 species a year. Paul Ehrlich and Thomas Lovejoy echoed similarly gloomy predictions. By 1992, Wilson was speculating that extinction rates might be between 27,000 and 100,000 species per year (280), though in recent times these alarming estimates have come under considerable criticism, particularly in Bjørn Lomborg's *The Skeptical Environmentalist* (2001, 249–57).[1]

Wilson's was the first of a number of books that have popularized the term "biodiversity." "Biodiversity" is, of course, a blend of the phrase "biological diversity." The term was coined in 1985 by Walter G. Rosen for "The National Forum on BioDiversity," a conference held in Washington DC in 1986 (Harper and Hawksworth 1995). Its proceedings were edited by Wilson (1988) under the title *Biodiversity*. While Wilson was certainly concerned with species numbers, an important theme of his work, and one that we shall echo, is the idea that diversity cannot be captured by species numbers alone. Species are important, for at least in eukaryotic organisms biological diversity is parceled up into independently evolving lineages. But we cannot assume that we can measure the diversity of a system just by counting species numbers. For the functioning of biological systems—as one quote with which we begin reminds us—depends on the kinds and combinations of organisms present, a fact whose importance has recently been reiterated by Kevin McCann (2007).

However, if we accept that biodiversity is important, and if we accept that there is more to biodiversity than species number, we must be able to answer the following questions: What is diversity and how do we measure it? What is the appropriate focus for conservation biology? We shall see that, from the beginning, there has been a potentially troubling ambiguity in thinking about biodiversity in conservation biology (and hence applied ecology). The ambiguity is between what conservation biology wanted to conserve and the mechanisms of conservation. Biodiversity is sometimes thought of as a measure of what we want to keep, but it is sometimes also thought of as a tool: a measure of an instrumentally important dimension of biological systems. So in much of this book we will address the purposes to which the idea of biodiversity is put. Disciplines are shaped by their histories, which transmit legacies of tools, assumptions, and projects from one generation to the next. We begin by asking why ecologists and conservation biologists started using the term.

Conservation biology is a young discipline because the idea that we ought to put scientific and governmental resources into conserving nature is a relatively new one. In the United States in 1873 it was suggested that the government ought to "preserve spots of primitive land-surface of which the vegetation was especially interesting" (Bocking 1997, 21), and indeed there were some limited successes in this respect early in the twentieth century. However, it was not until after the extremely intensive land use of the Second World War that governments began to take seriously the idea that wilderness areas were in real danger of disappearing.[2]

From its early beginnings, the conservation movement had been tied to the developing science of ecology through the work of pioneers such as Arthur Tansley, who was professor of botany at Oxford from 1927. One of the problems with integrating the conservation movement with the science of ecology was providing a clear characterization of the target of conservation. Ideas such as "wilderness" (let alone "interesting vegetation") are at best vague and imprecise. Worse, they often rest on confused concepts of the natural (as if humans were not part of nature), and a naive view of the extent of human influence on the biological world. If a wilderness is an ecological system unaffected by human activities, there are none. There are no untouched deserts or pristine tropical rainforests. So despite the fact there is more to biological diversity than species numbers, it is not surprising that in the 1960s and 1970s the focus changed to the preservation of species. The great advantage of this formulation of the agenda of conservation biology is that we can identify species and their extinction more reliably than we can identify a wilderness and its domestication (despite prolonged debate concerning species definitions; see chapter 2). Switching attention to species meant that conservation could be treated as a scientific enterprise. Moreover (and this will be an important theme of our whole book), perhaps species richness is a good index of conservation ends more generally: in conserving one, we may conserve the others.

While in principle almost all conservation biologists think there is more to biodiversity than species richness, in practice some measure of species richness is typically used in conservation planning. That is especially so because switching focus to species also meant that conservation principles could more easily be passed into law. This began in the United States with the passing of the Endangered Species Preservation Act (ESPA) in 1966. That law authorized the secretary of the interior to make a list of endangered native fish and wildlife and "insofar as is practicable" to direct federal agencies to protect those species. The ESPA was superseded in 1969 by the Endangered Species Conservation Act and then again in 1973 by the Endangered Species Act. With each iteration, the legislation was strengthened until finally it was made illegal to kill, harm, or otherwise "take" a listed species. At this point, despite its obvious advantages, the new conservation strategy was clearly going to lead to controversy. Push finally came to shove in 1973, with the snail darter.

In 1973 researchers discovered a new species of minnow living on the gravel shoals of the Little Tennessee River. The snail darter (*Percina tanasi*) is ten centimeters long (fig. 1.1). Even by the unexacting standards of minnows, it is unremarkable. However, as this stretch of river seemed to be its only natural habitat and as the Tennessee Valley Authority was constructing the $116 million Tellico Dam on the same stretch of river, the snail darter was duly placed on the endangered species list. The ensuing legal battle over the fate of the beleaguered snail darter was both prolonged and acrimonious. In 1978 the Supreme Court halted the construction of the Tellico Dam, which had already cost $78 million. Later that year Congress responded by creating a committee (immediately dubbed the "God squad") that had the power to exempt selected species from protection. However, in 1979, at its first meeting, the committee ruled that the snail darter should indeed take precedence over the Tellico Dam. The final act in this drama came in 1979, when the Tennessee congressional delegation slipped a rider into an appropriations bill exempting the Tellico Dam from the Endangered Species Act. It passed narrowly and the dam was built. In a final twist, in 1980 other wild populations of the fish were discovered; it was not endangered after all (for a much more comprehensive history of this controversy, see Nash 1990).

Herein lies the problem. On the one hand, environmentalists were determined that the U.S. government ought not knowingly let an endangered species go extinct for the sake of economic benefit. On the other hand, the proponents of the dam were equally adamant that thousands

FIGURE 1.1. The beleaguered snail darter (*Percina tanasi*). Image courtesy of U.S. Fish and Wildlife Service.

of jobs should not be put at risk for the protection of an undistinguished species. They argued that:

- There are many species of minnow, thus the snail darter was phylogenetically uninteresting.
- The snail darter was not phenotypically distinctive.
- The snail darter had no economic importance and, as it had only recently been discovered, there were no important cultural traditions associated with it.
- The snail darter was a species with a small population and limited distribution. Its extinction was unlikely to have flow-on effects on the biota at large.

Americans care about endangered species. It was a strong and persistent public outcry about the fate of the whooping crane that caused the first piece of endangered species legislation to be passed in 1966. But the snail darter saga tells us that most Americans (and we are willing to bet, most people everywhere else) don't care about all endangered species equally. And surely this discrimination is rational. From the point of view of conservation biology, the argument that seems to have carried the day was the claim that the snail darter was just not sufficiently distinctive, or as we might say now, that the loss of the snail darter did not constitute the loss of significant biological diversity. So even if we take the aim of conservation biology to be that of conserving species rather than biodiversity conceived more broadly, we are still faced with the problem of ranking. Which species should be conserved and at what cost?

1.2 BIODIVERSITY AND BIODIVERSITIES

As the historical excursion above will have made clear, the concept of biodiversity was coined at the intersection of science, applied science, and politics. Moreover, though most who talk about biodiversity think that there is something important about it, there are very different rationales for its preservation. Thus, some have argued that biodiversity ought to be conserved because it is a feature of the natural world that people enjoy and find useful. It has what conservation ethicists call "demand value." It is a human end in itself. However, there are alternative, instrumental reasons for defending biodiversity. For example, an influential line of thought connects biodiversity to ecosystem function (see 6.4), and ecosystem function is of great economic importance. Such instrumental rationales for the preservation of biodiversity are complicated by our considerable ignorance about many threatened ecosystems and the biodiversity they protect. Most extant species are yet to be described by taxonomists (Holloway and Stork 1991), and we know very little about the distribution and abundance of most of those that have been described. Close to a million species of arthropods have been described, but we can assess the conservation status of only about 3,500 of them (Brooks et al. 2006). Thus there are many species threatened with extinction about which we know little, except perhaps that they have relatively small ranges and that they are unlikely to perform unique ecological functions. In this respect, the snail darter is typical. It is vulnerable to relatively local habitat change, and the extinctions of species vulnerable for those reasons are unlikely to have dramatic flow-on effects (with the possible exception of island species, all of whom have restricted ranges). That in turn makes it unlikely that there are powerful economic-instrumental reasons for preserving such species. Perhaps we should think of these unremarkable species as expendable (see Sober 1986). However, there is a precautionary principle to be urged against this thought: if we let a species go extinct, we have foreclosed on the possibility that we might discover the species to be important. We ought to preserve biodiversity to hedge our bets. We maximize what conservation ethicists call "option value." These ideas will be explored in detail in 8.3 and 8.4.

Since the concept of biodiversity has been forged from such different sources and with such different motives, it is no surprise, then, that it has been used and measured in widely varying ways. We will mostly focus on the idea that biodiversity is a natural magnitude (or magnitudes) of biological systems, for this is often how biologists employ the concept (Gaston 1996a, 1996b; Kinzig et al. 2001). Indeed, biodiversity is often spoken of as if it were a single property, something that we might measure and compare across two habitats (Rolston 2001), and this idea continues to be influential in conservation biology, though conservation biologists no longer expect to be able to measure biodiversity directly. As we shall see in chapter 7, there is considerable discussion in conservation biology about surrogates, readily identifiable and measurable features of biological systems. According to those searching for a surrogate, biodiversity itself is a complex property, but if we are lucky it covaries in a reasonably reliable way with a simple and measurable property. We need a measure of relative importance, change over time, and of the effectiveness of intervention. So surrogates are chosen as biodiversity indexes: we can use them to measure the biodiversity difference between habitat patches at a time, thus setting relative conservation priorities. And we can use them to measure biodiversity changes over time, thus alerting us to troubling changes, and enabling us to evaluate the success or failure of protection. But these surrogates are supposed to be measures of overall richness or variety.

For example, in their influential overview of conservation planning, C. R. Margules and R. L. Pressey write:

> Biological systems are organized hierarchically from the molecular to the ecosystem level. Logical classes such as individuals, populations, species, communities and ecosystems are heterogeneous. Each member of each class can be distinguished from every other member. It is not even possible to enumerate all of the species of any one area, let alone the members of logical classes at lower levels such as populations of individuals. Yet this is biodiversity, and maintaining that complexity is the goal of conservation planning. (Margules and Pressey 2000, 245)

They are not alone. In a similarly wide-ranging and much cited overview, Craig Groves and his colleagues define biodiversity as "the variety of living organisms; the biological complexes in which they occur, and the ways in which they interact with each other and the physical environment . . . this definition . . . characterises biodiversity as having three primary components, composition, structure and function" (Groves et al. 2002, 500).

We will be interested in this idea of biodiversity as a natural feature of biological systems, though like Kevin Gaston and John Spicer (2004), we will reject the idea that there is a single measure of the diversity of a biological system. We doubt, in fact, that anyone really thinks there is a single natural property of a biological system that captures all its biologically relevant diversity, though perhaps Daniel Brooks and Deborah McLennan come close, suggesting that diversity is essentially species in their phylogenetic structure. They begin their 1991 monograph with a thought experiment about a tidal pool, inviting their readers to compare how much they know about an organism in the pool if given ecological information (the organism is a predator) or if given phylogenetic information (the organism is a fish). A predator, after all, might be an octopus, a starfish, a crab, or a fish, yet a starfish and an octopus differ far more than any two fishes (Brooks and McLennan 1991; 2002).

We will not find much reason to accept the idea that diversity is essentially captured by species and their phylogeny. But we shall see that a somewhat more modest view deserves to be taken seriously: that a phylogenetically informed species count is a good general purpose indicator or surrogate for total biodiversity (see, for example, Forest et al. 2007). We discuss a number of proposals for meshing species richness and phylogenetic information in section 7.3. A more pluralist position insists that there are distinct dimensions of biodiversity, and the form of biodiversity of interest depends on what a biologist wants to understand about the system in question. If, for example, we are interested in the stability with which a given region provides ecosystem services, we need to identify the ecological roles of the organisms in the system: the organisms that fix nitrogen, the detrivores that recycle dead plant matter, the pollinators, and so forth (the "guilds," as they are sometimes known). In contrast, if we want to know whether the species structure of the system is stable, then we will need to identify whether (for example) those constituent species are divided into populations that can rescue one another by migration. In other words, if we change focal question, within-species diversity becomes as important as between-species diversity.

In thinking about whether there is a single specification of biodiversity, it is important to distinguish between two forms of pluralism. On one version, the same system can be analyzed differently, depending on the predictive and explanatory purposes of the investigation. Thus we have suggested that if we were interested in ecosystem services we might measure very different properties of a system than if we were interested in the stability of taxonomic composition. In both cases, the variety or differentiation within the system is salient, but the elements over which variation is defined will be quite different. A second version of pluralism suggests that different systems need to be analyzed differently. For example, we might think that species richness is a good measure of the biodiversity of marine invertebrate communities but not of microbial communities. We will be mostly, but not solely, interested in the idea that we need to identify diversity differently, for different explanatory projects. Of course, even if that is true, there might still be important causal relations between these dimensions of diversity, such that each might partially predict the other.

The discussion above suggests that biodiversity might not be a single natural property or quantity: that biological systems are biodiverse in more ways that one. Sahotra Sarkar is even more skeptical: he uses "biodiversity" to mean roughly whatever we think is valuable about a biological system. That would make biodiversity as varied and valuable as human tastes and goals (Sarkar 2002). In his 2005 work, Sarkar continues to be skeptical about the prospects for defining biodiversity, arguing that evolutionary and ecological taxonomies are themselves imperfectly defined, and that each is partially incommensurable with the other. Even the conservation of genes, species, and communities would not, he argues, conserve all biologically interesting phenomena, for some result from unique interactions between proper components of these systems (Sarkar 2005, 179–82). Our main focus will be in the middle ranges of the spectrum of views from Holmes Rolston to Sarkar. As we have just noted, we doubt that there is a single natural property that captures the total diversity of a biological system. But neither do we think that the gastronomic or medico-herbal biodiversity of a rainforest has the same status as an account of its species richness.

1.3 HISTORY AND TAXONOMY

Assessing the biodiversity of biological systems—a coral reef, tropical rainforests considered collectively, the entire biota at some point in time—depends on recognizing the atoms in that system. This most often takes the form of an inventory constructed using a classification system: a way of recognizing the significant elements in that system, and a specification of their important similarities and differences. Quantification involves counting. But we cannot just count; we must count *something*. We must be able to say "Another one of those"; but to what does "those" refer? In general, the diversity of a system will depend both on the number of distinct elements in the system and on their degree of differentiation. Once we know what to count and how to compare, we can take both factors into account in a conceptualization of biodiversity, and we can ask whether and why diversity, so conceptualized, matters.

In this section, we discuss the general problem of classification systems in biology, taking as our stalking horse the most familiar example: the Linnaean classification system, the system that begins by classifying species into genera, that is, into sets of closely related and similar species. This system is not just the best-known classification system in biology; it is also of fundamental importance given the common practice within conservation biology of using species and species richness as proxy for biodiversity in general.[3]

Natural Classification

To understand a system we need to identify the units out of which the system is built, and whose actions and interactions drive the system. And we have to identify the crucial differences between those units. This is true of biological systems, but not only biological systems. Thus in trying to understand human cultures we need to identify the agents whose interactions constitute those cultures. Are all social agents individual human beings? Or do they include certain collective agents as well (tribes, firms, unions, and other institutions)? Moreover, we have to identify the crucial similarities and differences between human agents. Understood this way, constructing a classification system is far from trivial. Solving problems of this form was the key to the revolution in understanding chemical systems that began in the late eighteenth century. Indeed, solving the units-and-differences problem is central to any attempt to understand a domain. Moreover, a good taxonomy is an enormously important tool, because a good system of classification links diagnostic criteria for identification with similarity in causal profile.

Consider, for example, the folk psychological category of anger. Anger is diagnosable; it is not difficult to recognize the truly angry. And anger is causally significant; the angry are disposed to act in rather similar ways. Classification systems that combine recognition criteria with causal salience in this way are known as *natural classification systems*, and they code information effectively. A natural system links together individuals with similar causal attributes, so when we recognize a further member of the same category ("another angry motorist," we think warily) we have a lot of information about the likely behavior of that new individual. A natural classification system in biology has the same advantages. To say that two organisms are members of the same taxonomic group is to say that they are importantly similar and, depending on the taxonomic system employed, to license inferences based on those similarities. Importantly, they will be similar with respect to features whose existence or importance we are yet to discover. The great strength of a good system of taxonomic classification is that it allows us to infer a great array of facts about the physiology, ecology, and behavior of a specimen based upon common features of the better-known members of the taxa to which it also belongs. The enormous scientific effort expended on understanding the developmental biology of just a few model organisms (a fruit fly, a nematode worm, a mouse, a fish) is based on this intellectual strategy.

In contrast, astrological categories are not natural. One of us (Sterelny) has the same astrological sign as his daughter: we are both Scorpios. This characteristic is diagnosable; it is not hard to identify further Scorpios. But such identification tells us nothing, for Scorpios lack a common causal profile. Race-based classification of other humans is probably intermediate between the truly hopeless astrological classifications and genuinely natural ones. Our ordinary ethnic categories only map very roughly onto biologically distinct human populations (Cavalli-Sforza et al. 1994). Moreover, people are often apt to infer too much from putative membership of such categories.

Thus taxonomies are important cognitive tools, and hence we are rightly concerned when our categorizations go awry. Developing a natural classification system is often a major intellectual achievement. The claim that a system in use is not natural, likewise, is an important intellectual challenge. So changes in the way we view biological taxonomy are not just changes in scientific fashion. They are changes in our view of the units-and-differences project, and that is a fundamental project in biology.

However, the problem of constructing a natural system is especially difficult for biology. Identifying similarity and difference is difficult because much of biology is profoundly historical. It is historical not just because (some) biologists aim to chart and explain a particular historical process—the evolutionary history of life on earth—but because biological systems—organisms, populations, gene pools, species, communities, ecosystems—are the products of historical processes. And biological systems differ from one another in part as the result of those historical processes (Williams 1992). That is not true of physical and chemical kinds.[4] Gold has a history; all gold is made in stellar explosions. But the different tracks particles of gold have made through time and space make no difference to their intrinsic causal profile. No one wonders whether the ductility, reactivity, or melting point of gold will be different on the planets of other solar systems. In contrast, in biology history leaves its traces on organisms. The desert-adapted flora and fauna of Australia resembles the arid-lands biota of Africa in some respects because of their similar environments, but the biotas differ importantly because of their different pasts.

Organisms (populations, species) are the result of a conspiracy between history, environment, and chance. Since those conspirators mark biological systems in different ways—affect their causal profile in different ways—it turns out that there is no single system for identifying all the similarities and differences between biological systems that matter. Nothing in biology is exactly equivalent to chemistry's periodic table or to geologists' classifications of minerals. This makes a profound difference; as a consequence, there is no single right way of identifying the elements of biological populations or of identifying the differences between them that matter. Developing this case and assessing its consequences will be the burden of this whole book. But we begin the argument with an illustrative case, by sketching a brief history of the taxonomy of species and species differences. This history is in many ways an attempt to build a taxonomy that recognizes and integrates shared histories with phenotype similarities. No stable solution has been found. Current practice sacrifices phenotype similarity and uses shared history as its basis for identifying taxa. But that is not because phenotypes are unimportant.

The search for natural classification systems in biology has turned out to be difficult, because biological individuals are marked by both their history and their environment. It has proved to be difficult (arguably impossible) to incorporate both influences on causal profiles within the one system of classification. If there were a single natural taxonomy for biology, the biodiversity problem would be more tractable. We could unequivocally identify the natural elements from which biological systems are composed, and their important similarities and differences. Defining diversity would still not be easy; some systems would be diverse because of the number of distinct elements in them and others because of the differences between those elements, and so we would have to weight differentiation against number. But as we shall see, the quest for natural taxonomies in biology has been difficult, and that exacerbates the problem of defining diversity.

Evolutionary Taxonomy's Uneasy Compromise

By the publication of Darwin's *On the Origin of Species*, in 1859, the Linnaean system was already in wide use in biology. The basic Linnaean move was to introduce the binomial system, with species being grouped into genera, each of which consists of a cluster of similar species. But it was elaborated into a deeper hierarchical system: a cluster of similar genera is a family; a cluster of families is an order, and so on up the taxonomic hierarchy. This system was one of several nineteenth-century systems of taxonomy based upon elaborate patterns that, given a certain amount of charity, were there to be found in nature. But these systems lacked any explanation of the patterns on which they were based.[5] Darwin changed all that. The idea behind evolutionary taxonomy was that if evolution was the process responsible for natural variety then a taxonomic system based on the historical relationships between species promised to be both fundamental and predictive. Fundamental because evolution was the shaper of living things. Predictive because if most characters turned out to be inherited then genealogical proximity would predict phenotypic similarity. So this taxonomy makes an empirical wager that taxonomy based on phylogeny will be predictive, stable, and explanatory. The Linnaean system was given a historical reinterpretation in terms of common descent. A genus is a cluster of species whose common ancestor lived relatively recently. More inclusive taxonomic ranks (families, orders, hierarchies), likewise, are groups related by a common ancestor, but with the joint ancestor deeper and deeper in time. The first horse, the founding species of the family Equidae, is much more ancient than the first member of *Equus*, the surviving horse genus.

So, the idea was simple—taxonomy would reflect similarity with respect to inherited characters.[6] But it was also supposed to reflect the extent of phenotypic divergence. High taxonomic ranks reflect not just depth in time but also the extent of divergence. Thus a *sufficiently distinctive species* (like the kakapo or the New Caledonian kagu, the sole species in its family) might warrant its own genus. However, since all of the species in (say) a given genus are supposed to be close relatives, characters used in classification should reflect evolutionary history. "Homologies," as they are known, are character states that have evolved only once in a given lineage and have subsequently been inherited. Groups that share many homologies are closely related, and these closely related groups form the genera, families, classes, and so forth of the new taxonomy. If two assumptions are satisfied, the resulting taxonomy will be both explanatory and predictive. But phenotypic similarity and genealogical relationship must covary, and it must be possible to recognize evolutionary relationships reliably by identifying homologies.

Evolutionary taxonomy ran into trouble, as both assumptions proved less tractable than had been anticipated. Evolutionary taxonomists expected that evolutionary theory would identify the traits likely to evolve only once. This expectation proved to be too optimistic. For example, one idea was that characters occurring across a large range of environments (called "broadly adaptive characters") would evolve much more slowly than specialized characters. This slower rate of change implies that broadly adaptive characters would reflect evolutionary history; shared broadly adaptive characters are probably homologies. Thus, gnawing in rodents and flight in birds seem to be constant across a great range of environments, so these characters probably pick out natural groups that reflect evolutionary history. However, not all widespread adaptations are homologies. Vision in insects and vision in mammals occur across a broad range of environments, but insects and mammals do not share a common, sighted ancestor. Moreover (and perhaps more importantly), the evolutionary theories that guided us in identifying probable homologies were often supported by claims about the evolutionary history of particular lineages. Since evolutionary taxonomists constructed their phylogenies on the basis of quite controversial aspects of evolutionary theory, and yet supported these by phylogenetic claims, the critics of evolutionary taxonomy suspected that this method of identifying homologies was circular.

Evolutionary taxonomists argued that experienced taxonomists were often successful at picking out natural groups, but, inevitably, evolutionary taxonomy suffered from the suspicion that "homology detection" was at best an inexact science. Moreover, while evolutionary taxonomy was a system based on evolutionary relationships, it was not based *only* on evolutionary relationships. It also respected the degree of evolutionary change in a lineage. As evolutionary taxonomists used the term, "dinosaurs" did not include the living birds, because though they descend from dinosaurs they are too different from the characteristic, definitional dinosaurs. Likewise, "reptile" included the ancestral, ancient reptiles plus some but not all of their descendants. It includes the crocodiles, tuataras, snakes, lizards, and turtles, but not the mammals and birds. Only the exothermal, egg-laying, scaly descendants are reptiles, because they alone sufficiently resemble the ancestral reptiles. But how similar is similar enough? How different is too different? Suspicion about these aspects of evolutionary taxonomy gave rise to the phenetic revolution.

Similarity Is Not Enough

In a paper written in 1940, the botanist J. S. L. Gilmour argued that taxonomy should not attempt to represent diversity in a way that reflected evolutionary history. He was worried by the circularity problem we noted above, arguing that we should stop trying to identify some particular subset of characters whose special status would underpin the taxonomic system. A classification should be based on all the attributes of the individuals under consideration (Gilmour 1940, 472). The aim was to base classification objectively, on overall similarity, rather than relying on intuitive or theoretical guesses about the importance of some characteristics and the unimportance of others. The overall similarity of groups of organisms would be calculated by summing the similarities of as many characters as could practicably be measured (see, for example, Sokal 1985). The resulting theory has come to be known as "phenetics."[7]

However, the project of building a classification system based on overall similarity is hopeless. If any characteristic at all counts in determining similarity relations among (say) a house fly, a fruit fly, and a bee, then they are all equally similar and equally unlike one another. For every individual has, and lacks, an infinity of characteristics. Almost all of these are of no interest at all. For example, the organisms that compose these three taxa will vary in their average distance from Britney Spears's navel the instant she turned eighteen. But this property is of no interest to invertebrate biology. Overall similarity is not a well-defined concept, as Nelson Goodman vigorously remarked in "Seven Strictures on Similarity" (1972, 437). So phenetics in particular, and biological taxonomy in general, needs a principled solution to the problem of identifying the traits to measure and compare. No such solution can be theoretically neutral (as the pheneticists had hoped). For the identification of characteristics will depend on their importance to biological processes.

Moreover, there is a second problem for the pheneticist. Once groups of organisms have been described in terms of their character states, a variety of statistical methods can be used to produce a measure of the "phenetic distance" between them. The following decades saw the development of many algorithms (known as ordination methods) whose aim was to produce natural groupings or clusters from phenetic distance data. But rather than discovering an ideal ordination method, the investigation just seemed to produce a proliferation of possible methods, all of which had their adherents. This problem has been the source of long-standing criticism. The following quote from Mark Ridley (1986, 164) is forthright in tone. It is therefore not representative of a debate that is widely recognized (Hull 1988) as being downright acrimonious:

> The aggregate phenotypic similarity among a pair of species depends on the statistic used to measure it: it has no objective, natural existence. There are many measures of it, and they give different classifications. The phenetic taxonomist has to choose among them. This choice is subjective. Although the procedure, once a statistic has been chosen, is repeatable, the choice is subjective.

The patterns of similarity and dissimilarity between species are certainly an important aspect of biological diversity. But phenetics was not capable of giving an account of the theoretical foundations of those patterns.

The Triumph of History

Cladism achieves objectivity by choosing a system of classification that explicitly represents only the historical connections among species. This system was due largely to the work of a German entomologist named Willi Hennig. His classification system is based solely on the detection and representation of evolutionary history. Evolution by speciation causes living organisms to be related by the treelike hierarchy known as a phylogeny. The real groups of which species are part are monophyletic lineages of species. A monophyletic group or clade is any

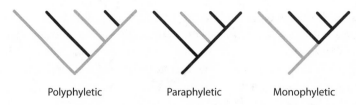

Polyphyletic   Paraphyletic   Monophyletic

FIGURE 1.2. Three types of classificatory grouping. The nodes represent speciation events. The solid lines represent the species that belong to each of the
groups.

branch of a phylogenetic tree that includes an ancestral species and all
and only the descendants of that species. A paraphyletic group is one
that includes an ancestral species and some of the descendants of that
species (see fig. 1.2).

On this system, similarity and difference between species is just genealogical distance. The bonobo and the common chimp are maximally
similar because they are sister species: they are the only extant descendants of their most recent common ancestor. Each is equally closely
related to our species, and more closely related to us than to any other
living species, because we are only one branching event more distant.
Once there lived a species that was our ancestor, the ancestor of both
chimp species, and the ancestor to no other living species (fig. 1.3).

Clades are well-defined and objective chunks of the tree of life, but
they do not represent phenotypic diversity explicitly. Phylogenetic
structure might be a reasonable guide to phenotypic divergence, but
representing such divergence is not part of the task of systematics. Hennig rejected the explicit representation of phenotypic diversity (just as
the pheneticists had rejected the explicit representation of phylogeny).
That is one key difference between cladism and evolutionary taxonomy.
The other was a method for recovering genealogical relationships from
biological data that makes conservative, uncontroversial assumptions
about evolutionary mechanisms. For Hennig and the cladists who fol-

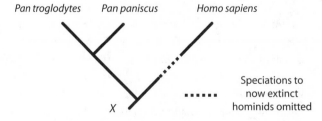

Pan troglodytes   *Pan paniscus*   *Homo sapiens*

Speciations to
now extinct
hominids omitted

X

FIGURE 1.3. Recent hominid phylogeny. *X* marks the unique species ancestral to
bonobos, common chimps, and humans.

lowed him, the only characters that matter in identifying genealogical relationships are shared derived characteristics. Marsupials, for example, have many of their similarities not in virtue of being *marsupials* but in virtue of their membership of the larger clade of the *mammals*: most obviously their fur and the capacity of females to lactate. These are inheritances derived from a deeper ancestor than the Mother-of-all-Marsupials. Hence they tell us nothing about relationships within the mammal clade; a character trait that evolves before a clade splits from its ancestral stock cannot carry information about relationships within that clade (though subsequently evolving modifications might do so).

This is a conceptual point; it does not depend on controversial claims about evolutionary mechanisms. We illustrate it with a few antipodean examples. The marsupials' pouch, together with various aspects of their dentition and physiology, *are* inheritances from the Marsupial Mother, and hence those traits are informative about relationships within the mammal clade. They are shared and derived. They evolved within the mammal clade (they are derived) and they are shared across the species descending from their point of origin in the mammal tree (hence they are shared). Marsupials' pouches mostly open toward the front of the animal. Thus when a female kangaroo is at rest, the pouch opens upward, and her joey is in no danger of falling out. But not all marsupials have front-opening pouches. The opening of a wombat's pouch is posterior rather than anterior (otherwise it would tend to fill with dirt as the wombat burrowed into the earth). This character is shared and derived within the marsupials, and hence is evidence supporting the genealogical proximity of the common wombat and the southern and northern hairy-nose wombats. The pointy ears of the two hairy-nosed species is a derived character that supports their status as sister taxa, more closed related to each other than either is to the common wombat.

Unlike evolutionary taxonomists, cladists did not expect shared derived similarities between organisms to have special embryological or ecological markers. Instead, they proposed to rely on the idea that similarities due to convergent and parallel evolution would be rare compared to similarities due to inheritance. Change is rare compared to nonchange. This is an empirical but relatively uncontroversial claim about evolutionary processes. On the basis of these assumptions, cladists take it that phylogenetic hypotheses that minimize the number of changes needed to account for observed patterns of similarity and difference have the best chance of being right. This method of detecting phylogeny is known as parsimony analysis.[8] A phylogenetic hypothesis that minimizes the number of character state changes among (say) the

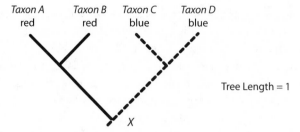

FIGURE 1.4. The most parsimonious phylogeny, said to have a tree length of 1 as it depicts one change in a character state.

marsupials is more likely to approximate a representation of mammalian evolutionary history than a less parsimonious hypothesis.

The basic idea is simple. If an ancestral species has blue plumage and so do its modern descendants, then it is likely that they have retained their ancestral color (as in fig. 1.4). The less likely alternative is that some have, in the interim, changed their color and then changed back (as in fig. 1.5). By comparing the histories of character state changes in closely related species, cladistics allows us to detect phylogenetic structure and thus to represent it taxonomically. In practice, of course, it never goes this smoothly. There are reversals and convergences, and often the most parsimonious tree constructed from one character set is not identical to that using other characters. As a result of these reversals and convergences, parsimony analysis typically results not in a single most likely phylogeny, but in many roughly equally likely phylogenies.

Cladistics has been very successful in recent decades. However, it is not without its problems. It is vulnerable to its empirical assumptions about evolutionary process. For example, cladism (in its usual form) assumes that in branching a species splits into descendants that split again; they do not fuse. While this is a plausible view of the evolution of multicellular animals, it may not be a universal feature of life, especially not prokaryote life (see O'Malley and Dupré 2007; Goldenfeld

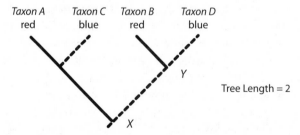

FIGURE 1.5. A less parsimonious phylogeny, this time with a tree length of 2.

and Woese 2007). Microbial taxonomists have traditionally been much more concerned with phenotypic characters than with phylogenetic relationships (Goodfellow et al. 1997, 26). This is due in part to conjugation, a process in which bacteria are able to pass genetic material (and thus ultimately phenotypic traits) to other individuals that are not their own offspring. Such "cross-borrowing" tends to decrease the explanatory utility of cladistic analysis to the extent that these represent lineages of evolving populations of organisms.

There are many cladistic analyses of prokaryotic life; indeed, the idea that the tree of life is organized into three ancient domains, the archaea, the eubacteria, and the eukaryotes, is based on the discovery of an ancient split among the prokaryotes. But there is an important sense in which these trees are histories of gene lineages rather than organisms.[9] For one thing, these phylogenies are based on genetic data: the comparison of homologous genes. But more importantly, if cross-borrowing is a regular feature of bacterial life, we cannot assume that closely related genes—genes dating back to a recent common ancestor—are parts of the genome of closely related organisms. Among the eukaryotes, gene histories and organism histories are typically (though not universally) concordant. When two versions of a gene complex begin accumulating changes independently of each other, it will be because those genes' lineages are contributing to two organism lineages evolving independently of each other. If genes are often transferred laterally, that is an assumption we can no longer make. We cannot treat the branching pattern of gene evolution as a proxy for the branching pattern of the organism lineages of which they are a part.

Moreover, there is a set of technical challenges to the tractability of their approach. Parsimony analysis is computationally intensive. Finding the most likely phylogeny is made difficult by the fact that the number of possible "trees" increases exponentially as more taxa are added to the analysis. Three taxa can be arranged in only three different rooted tree topologies (ones that show both phylogenetic relationships between taxa and also pick out a particular taxon as being the ancestor of all the others). Five can be arranged in 105. Ten can be arranged in 3.4×10^7, while twenty can be arranged in 8.2×10^{21} (Quicke 1993, 59). It is true that there are algorithms to find the most parsimonious tree that do not involve exhaustively searching all the possible tree topologies. Even so, in practice computation tractability will always be an issue for cladistic analysis.

In short, cladistics is explanatorily powerful, with a well-defined rationale. These are great virtues in a taxonomic system. But it purchases these virtues at the cost of abandoning an explicit representation of phenotypic diversity. So while cladism might give a good account of the units out of which biological systems are composed, it is less plausible as an account of the relevant similarities and differences. Is that cost too high? The answer to this question depends on whether we think of cladism as a proposed solution to the problem of representing biodiversity. We think current cladistic taxonomy is best understood more modestly, as a response to *one biodiversity problem*: the representation of the diversity generated among the units of evolution. The cladistic solution to the units-and-differences problem is to identify the units as species and the difference as genealogical depth, while (very importantly) developing a methodology for making measuring and representing genealogical depth tractable. One important theme of much of this book will be an exploration of the extent to which biodiversity in this sense is a good surrogate or index of other aspects of biological diversity.

We return to the issue of the explicit representation of phenotypes, phenotype variation, and phenotype change in detail in chapters 3 and 4. This is crucial, since the variety of organisms—of phenotypes—is clearly central to biodiversity. We will argue in favor of the cladistic idea that there is no general metric of phenotypes through which we can compare the similarity and dissimilarity of organisms. But we argue in favor of a more restricted form of the explicit representation of phenotypes and phenotype difference. The phylogenetic information that cladistic methods provides enables us to represent phenotype differences as they evolve in specific lineages, to construct "local morphospaces," as we call them. This phylogenetic information enables us to make principled decisions about the traits we measure and compare. We develop this idea further in considering the relevance of modularity in chapter 5, in which we discuss development and diversity. These arguments become directly relevant to conservation biology in chapters 7 and 8, where we discuss attempts to combine phylogenetic information with measures of species richness. Since these attempts are based on cladistic theory, they too fail, and unnecessarily fail, to explicitly represent phenotypic information.

In the following chapters we will address other solutions to the units-and-differences problem, evaluating each as a tool for the representation and analysis of biodiversity. These need not be in competition with cladistic taxonomy. Ecologists, for example, sometimes talk about guilds or functional groups. Those units are not species; they are (often) phylogenetically diverse populations within an ecological system,[10] each of which has the same ecological role. For example, the pollinators in a woodland are a functional group, and they might include bees, birds, moths and butterflies, flies, bats, and possums (we discuss ecological taxonomies in chapter 6). An ecological taxonomy of functional groups within an ecosystem picks out new units and differences suited to the particular goals of ecology. But an ecological taxonomy of functional groups is compatible with a cladistic systematics of the taxa within them. Likewise, in chapters 3 and 4, we explore the idea that phylogeny identified by cladistic methods needs to be combined with a representation of phenotype evolution. Pluralism about biodiversity may be appropriate; as we have already noted, there may not be a single best representation of diversity.[11] That said, we would need to know how distinct proposals about the units-and-differences problem are related. When and how could such proposals complement one another; when are they in conflict? We begin to answer this crucial question about the relationship between different solutions in the next section. It turns out that there are two very different kinds of reasons for an interest in patterns in biological diversity: diversity can be important either as a cause or as an effect. To that idea, we now turn.

1.4 DIVERSITY AS CAUSE; DIVERSITY AS EFFECT

There is no theory-neutral characterization of the amount or kind of biodiversity in a biota at a time. For as we saw in 1.2, there is no solution to the units-and-differences problem without an account of the differences and similarities that are important. A judgment of importance is theoretically committing; it depends on a view of biological mechanisms and how they work. Theory choice also depends on the instrumental and explanatory purposes of particular groups of scientists. Thus if there are a number of complementary theories of a given biota it will follow that there are alternative, complementary specifications of units-and-differences. For example, community ecology, focusing on the array of populations in a habitat, and ecosystem ecology, focusing on the flows of materials and energy through a habitat, will typically describe the same biological system in quite different but apparently compatible ways. If biological theory is pluralist, relying on a number of complementary theoretical approaches to a given biological system, diversity in that system will be plural, too. But how plural, and in what ways? We begin by distinguishing between forward-looking and backward-looking theories of a biological system.

Conservation biologists are typically concerned with the *effects* of biodiversity and its loss. For example, they have typically argued that diversity adds redundancy and hence robustness at many biological scales. Genetically diverse species are buffered against environmental change; they are more robust. Arguably, biodiverse ecosystems are more stable, perhaps even more productive. It might even be that diverse global biotas hedge our bets against an uncertain future. But not all descriptions of biological systems are forward-looking in this way. In evolutionary theory, our interest in patterns in diversity is often motivated by the thought that differences in pattern are symptoms of differences in process: biodiversity patterns are informative signals of the processes that caused them. A good example of this is recent work on phenotypic diversity in evolutionary biology. Within microevolutionary studies, there is a long tradition that attempts to measure the strength of competition for resources between similar species by seeing whether competing species exhibit character displacement. For example, such studies measure whether two species of anolis lizards that live together on the same island are phenotypically different from populations of those same lizards when they are not in contact.[12] Divergence, if found, is a trace of competition. The same is true of macroevolutionary studies. For example, phenotype conservatism—no change over long periods of time—is often taken to be a signal of constraint on the power of selection to shape new forms of life. We will meet this interpretation of the evolution of the animals in 3.1.

Cladistic methodology is supposed to allow us to estimate the tree of life[13] while making only uncontentious assumptions about evolutionary mechanisms. Even so, the whole point of identifying genealogical relationships is to zero in on evolutionary mechanisms. We cannot estimate the extent to which Australian eucalypt phenotypes are adaptations to their current environment without a phylogeny that tells us which of their traits are shared derived inheritances from a pre-arid Australia, and which are convergent or parallel adaptations to their new and harder world. If eucalypts' characteristically hard, waxy leaves evolved before the great Australian drying, they cannot be an adaptation to that drying. Identifying phylogeny is essential to understanding phenotypic change.

Patterns in speciation are also signals of evolutionary process. Why are there so many beetle species? How was it possible for the cichlids in the east African lake systems to evolve so many species so fast? Patterns in the overall shape of the tree of life are signals of the processes that produce those patterns, and that is one reason why it's important to have a principled and objective characterization of those patterns. We need a well-established phylogeny to show a clade is unusually species rich or unusually morphologically diverse. Consider, for example, one of Wallace Arthur's striking examples of developmental constraint. As a group, centipedes vary considerably in segment number. But among the *Lithobiomorpha* centipedes there is no variation at all. All thousand or so species have their trunks divided into fifteen segments (Arthur 2000). Without a phylogeny showing that these species form a clade, this pattern of constrained variation is undetectable. Once that pattern is documented, it signals an explanatory problem. Cladistics really does capture an important aspect of biodiversity because it really does detect phylogenetic structure (albeit fallibly and sometimes with a great degree of difficulty). Phenotypes and phenotype differences can then be mapped onto that phylogenetic structure, revealing patterns in phenotype evolution. But notice the contrast between evolutionary and conservation biology. Systematicists have typically been interested in the mechanisms that cause (or fail to cause) diversity. Conservation biologists have typically been interested in the effects of biodiversity (and of its decline). So, for example, conservation biologists worry about the lack of genetic diversity in a species. Many New Zealand threatened species recovery programs are targeted on species (including the kakapo, black stilt, takahe, and perhaps some of the kiwi species) whose total populations are in the hundreds or less. Their effective breeding populations are of course smaller still, and that explains a concern about the *effects* of genetic diversity on extinction probability. These are models of the effects of (the lack of) diversity.[14]

One theme of the following chapters is that this distinction has been neglected in many theories of the nature of biological diversity. In arguing this, we take up and generalize a theme of Graeme Caughley's much-cited paper on the methodology of conservation biology (Caughley 1995). Caughley argued that (without noticing it) conservation biology had been working with two different and only partially compatible paradigms, the "declining population" paradigm and the "small population" paradigm. The declining population paradigm is backward-looking, as a declining population is a signature of unfavorable changes in the world of the organism in question. Hence, though backward-looking, it is a flag for action. The "small population" paradigm is forward-looking, for a small population is in itself a risk factor. Small populations are at inherent risk of extinction, from inbreeding, genetic drift, unpredictable external disturbance, and demographic stochasticity.

Naturally, a pattern of speciation or phenotypic variation across a set of related taxa might be both the causal signature of an important evolutionary event (an adaptive radiation, for example) and a causal input to downstream ecological and evolutionary events. The evolution of a set of specialists on an island archipelago changes those environments in important ways. Species richness can generate further species richness, as coevolving lineages show. Thus fig trees and fig wasps speciate together; speciation in one is matched by speciation in the other.

Even so, we cannot assume that a way of representing diversity that is optimal for the purposes of detecting some evolutionary processes is also optimal for explaining the input to others. For example, it might be important to identify cryptic sibling species in thinking about the effects of evolutionary processes; these might be markers of nonselective factors in speciation. But this distinction might not be important in characterizing the selective environments driving further evolutionary change. A more radical possibility is that in characterizing those environments we might need the ecologists' characterizations in terms of guilds or functional groups, rather than a genealogical specification of biological diversity.

These are very difficult empirical questions. But we cannot assume that a solution to the units-and-differences problem optimal for detecting the effects and relative importance of the different evolutionary mechanisms is also optimal for characterizing the environment in which ecological and evolutionary forces interact to generate further change. We think some attempts in conservation biology to incorporate phylogenetic distinctiveness into their metric of diversity do make just this assumption. For example, there have been recent defenses of the idea that conservation planning should give weight to phylogenetic distinctiveness, not just endemic species richness, on the grounds that in doing so we maximize the evolutionary potential of the diversity we conserve (see, for example, Mooers 2007; Forest et al. 2007). Phylogenetic distinctiveness is backward looking. On some ways of measuring it, we estimate the time since divergence from the common ancestors of the species in a region and sum those times. The total gives us a reading of the amount of evolutionary history those species represent. These methods of estimating importance heavily weight species (like the Tasmanian devil or the platypus) that have been long-separated from their nearest living relatives. But (as Mace et al. 2003 points out) species-poor lineages may be species poor precisely because they have little evolutionary potential. They have low intrinsic rates of speciation. They are "dead clades walking." If so, despite their high scores by these procedures, such lineages do not represent rich future possibilities at all.

1.5 PROSPECTUS: THE ROAD AHEAD

This book begins with conservation biology, and it will end with conservation biology; we return in the final chapters to the problems of both measuring and valuing biodiversity, with, we hope, a much richer understanding of the nature of biodiversity. In the road ahead, one theme will be central. To what extent is the species structure of a biota—its species richness in phylogenetic context—a good surrogate for bio-diversity in general? Is this structure a good way of identifying those patterns that are the signatures of ecological and evolutionary process, and a good way of specifying the input to further ecological and evolutionary changes? No one thinks that species structure is literally all there is to biodiversity. If the life sciences had perfect information about biological systems, and unlimited experimental and computational resources, biologists would not just count species. But the life sciences need a simple and empirically tractable model of total biodiversity. So perhaps the number and distribution of species, augmented in various ways for particular purposes, serves as a good multipurpose measure of local, regional, and global biodiversity. We begin this project in chapter 2, which seeks to resolve the somewhat puzzling fact that species richness is often used as a surrogate for overall biodiversity, even though the nature and identification of species continues to be controversial, and even though no one thinks that species richness is all there is to biodiversity. In chapters 3 and 4 we explore the relationship between species richness and phenotypic disparity, beginning with Stephen Jay Gould's well-known claims that species richness does not track disparity.

We extend the focus on phenotype diversity in chapter 5 by bringing development explicitly into the picture. Genes are paradigmatic developmental and evolutionary resources; the evolutionary plasticity and resilience of species depends to a considerable extent on the genetic resources available in species gene pools. So in that chapter, we discuss the diversity of developmental resources and explore the extent to which phylogenetically structured species richness is a surrogate for developmental diversity. In the final "filling" chapter of the conservation biology sandwich, we then turn to ecology. For us, the crucial question is whether communities or ecosystems function as biologically important organized systems. If they do not, if species (or populations) respond to environmental vectors independently of their neighbors' response, then species richness captures ecological diversity. Information about the species present, and the environmental variables acting on those species, would suffice for understanding ecological outcomes. That is not true if communities are organized systems. Ecosystem services, for example, would then depend on collective properties of the community. We then return, rather skeptically, to the problems of measurement and value as they have been conceived in conservation biology. Of course, conservation biology can be and is used purely as an instrumental, applied science to estimate specific dangers from threats to specific populations and to devise means of defusing those threats. Conservation biology can, and conservation biologists do, estimate the danger that (say) feral rabbits pose to breeding colonies of seabirds on Macquarie Island without taking any stand on the nature of biodiversity and its importance. But if we are right about biodiversity, conservation biology has yet to formulate its agenda coherently; it does not yet have a general and coherent account of what should be conserved and why.

2 *Species: A Modest Proposal*

Increasingly, agendas for future environmental research depend upon comparisons of estimates of species diversity. It is tacitly assumed that the units compared
are equivalent—an assumption that is clearly untenable when dealing with a diverse and unnatural assemblage like the algae. Despite this non-equivalence, such
comparisons continue to be made along with estimates by taxonomic specialists
in particular groups, of the numbers of species still to be described.

John and Maggs (1997, 84)

2.1 INTRODUCTION

We suggested in chapter 1 that the identification of biodiversity is tied to
the particular solutions to the units-and-differences problem that flow
from scientific theories, and that differing biological theories may not
identify the same set of units and the differences between them. Un-
fortunately, this pluralist possibility often goes unacknowledged. Many
of those studying biodiversity simply equate biodiversity with indexes
of diversity based on species richness (or with representation of higher
taxonomic categories). Thus much study of biodiversity assumes that
a unitary taxonomy provides a good representation of biodiversity for
most biological purposes, and that this unitary taxonomy is based on
species and higher taxa. That is puzzling given the longstanding disagreement about the nature or even the reality of species (Claridge et al.
1977). Species could hardly be a crucial component of biodiversity if our
species identifications reflect facts about human psychology rather than
the organization of the natural world (Hey 2001).

These disagreements about so-called species definitions have consequences for the measurement of biodiversity; they can lead to very

different views about the species richness of various clades. For example, Georgina Mace and her colleagues point out that the shift to a phylogenetic species concept (in which any taxon with a consistent, diagnosable difference from all others is recognized as a species) often leads to an explosion in the number of species recognized, especially in conjunction with genetic data. One example they give is of the addition of 140 new amphibian species to the Sri Lankan fauna. An impressive expansion, given that only 15 species had previously been recognized (Mace et al. 2003).

To make counting consistently across different systems even more difficult, every serious account of species recognizes intermediate cases. The famous biological species concept defines species in terms of reproductive isolation, but isolation comes in degrees (Ehrlich and Raven 1969). As most of us know, whether you are in with a chance depends on who else is at the party. New Zealand black stilts (*Himantopus novaezelandiae*) prefer to mate with members of their own species, but they will mate with pied stilts (*Himantopus leucocephalus*) if no black stilts are available. Instead of thinking that there is a class of individuals with whom some particular individual will breed, there is instead a breeding probability that decreases gradually as more and more inclusive groups are compared (Mishler and Donoghue 1982). The biological species concept must therefore recognize intermediate cases—populations that are neither conspecific with one another nor distinct species. So too must other species concepts (Sterelny and Griffiths 1999). For example, an alternative model characterizes species in terms of cohesion, but that is also a matter of degree. Indeed, it is most plausible to think of single populations rather than species as ecologically cohesive. This is because a species typically consists of a metapopulation, and the constituent populations of such species often occupy disparate habitats. A syngameon is a complex of genetically still connected but ecologically highly distinctive species (Seehausen 2004). These partially cohesive species (or semispecies as they are sometimes known) occur in both plants, such as white oaks and Pacific Coast irises (Arnold et al. 2004), and in animals, such as cichlids (Schliewen and Klee 2004). Syngamous clades are precisely those in which species cohesion is incomplete, and they are clades whose species richness is indeterminate.

So we begin this chapter by noticing two striking facts. First, in practice most explicit attempts to estimate biodiversity are attempts to estimate species richness. Second, evolutionary theory has been home to a long and continuing debate about the nature of species, a debate that has resulted in a profusion of species concepts. These disputes are not "merely semantic"; they reflect different views about the causes and consequences of differentiation between populations, of divergence and why it matters. One way of framing the topic of this chapter is: are there reasonable prospects for the development within biology of a consensus view of the nature of species? If so, would species richness then be a good general-purpose measure of biodiversity? We will see that there is a sensible motivation that warrants fixation on species. Species are empirically accessible. Phenomenological species are observable, identifiable, and reidentifiable aspects of the biological world. Moreover, phenomenological species correspond, in many cases, to evolutionarily significant lineages in the tree of life. So this approach to biodiversity does capture something real, despite the complexities of the species problem.

Even setting aside the complexities of competing species definitions, it is widely accepted that species richness does not capture everything central to biodiversity. Virtually all species-based accounts of biodiversity seek to represent not just the number of species in a biota but also their *structure* in some way. There are, however, very different proposals about structure. Especially within conservation biology, the measure of choice is often endemic species richness rather than species richness in general. Other proposals are ecological: most simply, assessing species abundance as well as species number. Other approaches include phylogenetic structure and/or phenotypic divergence with species richness. The simplest way of doing that is to keep track of the higher Linnaean categories represented in a biota: the orders, families, and classes, not just the species.[1] This simplest way is very common; the majority of biodiversity measurement strategies employ Linnaean taxonomy in some way or another. Many biologists think of biodiversity as taxonomic diversity (see, for example, *Species: The Units of Biodiversity* [Claridge et al. 1997]). Recent large-scale collaborative efforts to provide worldwide online databases of biodiversity have been primarily concerned with the collation and storage of Linnaean taxonomic information (for example, the Global Biodiversity Information Facility, www.gbif.org). In short, it is common ground that measures of species richness need to be supplemented. In this chapter, we set aside this complex of issues; they will be central to the following chapters. Instead, we concentrate on species richness itself. How can it be a core component of biodiversity, given the ongoing debate about the nature, and even the reality, of species?

This focus on species and how to define them in part reflects a shift in ambition in systematics. In chapter 1 we described the major taxonomic revolutions of the twentieth century. But alongside these major taxonomic revolutions there has been a more subtle change in biological systematics. At the opening of the twentieth century most taxonomists saw their jobs as roughly analogous to those of library cataloguers. Their aim was simply to produce a taxonomy into which all organisms could be placed and whose categories would be maximally informative and useful for practical purposes. If one thinks of taxonomy in this way, there is no fundamental difference between the levels of the Linnaean hierarchy. However, over the course of the twentieth century, systematics became a much more ambitious enterprise (Hull 1988). Systematists now "wanted their classifications to be more than just summaries of phenotypic variation. . . . Some wanted to discern entities that functioned in natural processes, particularly the evolutionary process" (Hull 2006, 796). Thus the species category was taken to be something like a natural kind (or, perhaps, several natural kinds that have been mistakenly lumped together).[2]

There are dissenters, but we think most of the participants in the great species debate have thought that the species category does pick out a natural kind. Hence there should be a causal profile that is common to each member of that category. Much of the species debate flows from different attempts to characterize the causal profile that is the signature of a true species. Indeed, there is a "species debate" only because biologists think species are a natural kind, or something like a natural kind. There is no "genus debate" or "family debate" because nobody thinks the same about genera or families.[3] We too think that the species category picks out something like a natural kind. In our view, both the focus on species and the profusion of species concepts reflects something real about and important about the biological world. The focus on species reflects the recognition that species are units of evolution and hence of biodiversity. The profusion of species concepts reflects the variety of species mechanisms and the complexity of the relationship between species, speciation, and the environment (though doubtless the different interests and backgrounds of biologists play some role in that profusion too). Despite those complexities, we shall also suggest that there are prospects for a limited consensus. For one important class of cases, the output of speciation mechanisms is a metapopulation that plays a distinctive evolutionary role. Species so characterized represent one significant form of biodiversity. In saying this, we have no intention to inflict upon the biological world another species definition. Rather, we intend to highlight some ideas common to a cluster of species concepts.

In the next section we confront the diversity of species concepts more seriously. That diversity, as we have remarked, flows in part from the profound biological differences between the different branches of the tree of life, and in turn those differences suggest that we have little chance of formulating a one-size-fits-all criterion that would allow us to recognize species across the different branches and thus enable us to measure the overall species richness of a region. One way of responding to the diversity of species concepts, then, is to conclude that the prospects for a species-richness based account of biodiversity are grim both practically and theoretically. They are grim practically because species lists are compiled using different species definitions, and these are not equivalent. They are grim theoretically because there is no single across-the-board criterion that we could use to make taxonomic databases consistent and well motivated. In 2.3, we argue that this is a much too pessimistic assessment of species-richness based accounts of biodiversity.

2.2 SPECIES, SPECIES CONCEPTS, AND SPECIATION

One response to the plethora of species definitions has been pluralism, the idea that there is no single, right species definition. However, pluralism in this context is worrying, for it seems to undermine the idea that species richness measures biodiversity. How could that be true unless we had an invariant species concept to use in counting biodiversity (see Mishler 1999, 313)? This is a legitimate concern, but we think it can be met. In 1.2 we distinguished between two forms of pluralism. Investigation-specific pluralists think that different theorists with their different explanatory agendas can legitimately describe one and the same biological system in quite different ways. A given population might be a valid species for a morphologist but not for a population geneticist. Philip Kitcher is a pluralist about species in this sense (see Kitcher 1984a; 1984b). Whatever its merits, this form of pluralism is no threat to the idea that we can compare the diversity of different biological systems by estimating their species richness. In contrast, system-specific pluralists think that different biological systems need to be characterized in quite different ways. Thus, for example, John Dupré (1993) argues that different types of organisms will be best classified using different criteria for specieshood. We think there are indeed crucial biological differences between the mechanisms that maintain the phenotypic and ecological integrity of lineages. As plenty of commentators have noted, the biological species concept, with its focus on barriers to gene flow between lineages, fits animals better than plants. So this form of pluralism does potentially challenge the idea that species are a common currency of biodiversity measurement. In section 2.3, we recognize the diversity of biological processes that cause lineages to split. Even so, we argue that a version of an evolutionary species concept can underpin a species-based account of biodiversity.

We have no intention of going through the many species definitions one by one (though the main contenders are outlined in Box 2.1). Those interested in such an analysis will find illuminating treatments in Claridge et al. (1997), Ereshefsky (2001), Wilson (1999), and Wheeler and Meier (2000). Rather, our aim is to explain why the debate has been difficult to resolve, and to contrast accounts that focus on the species-making mechanisms with those that focus on the evolutionary consequences of speciation. These downstream accounts, we will suggest, are more general and they do capture an important component of biodiversity. But this generality is at best partial. It captures macrobes not microbes, and, very likely, not all macrobes.

BOX 2.1: Some Important Species Concepts

Typological species

A group of organisms whose members sufficiently conform to a fixed set of characters. This is the "classical" concept of specieshood used by Linnaeus. Typology is the basis of species identification using keys (nested hierarchies of taxonomic characters that can be navigated so as to provide definite identification of a sample as belonging to a particular species). While practically useful, typology is fundamentally essentialist and thus rests on the false assumption that the identifying characters of species do not change over time.

Phenetic species

A group of organisms with a high degree of similarity with respect to a large number of taxonomic characters (Gilmour 1940; Sokal and Sneath 1963). Some problems with phenetic taxonomy were discussed in 1.2. The central issues to do with multivariate analysis apply to the species level as to any other level in phenetic taxonomy. There is no objective choice of similarity measure. Nor is there any justification within phenetics for the idea that the species level is more fundamental than other higher taxonomic ranks.

Biological species

A group of organisms that can potentially interbreed and that are reproductively isolated from other such groups (Mayr 1942). This definition obviously applies only to sexually reproducing species. Moreover, interbreeding potential comes in degrees. In his *Principles of Systematic Zoology* (1969), Ernst Mayr reduced this vagueness by removing the phrase "potentially interbreeding," though retaining "reproductive isolation." Hence the revised, less vague version of the definition comes at the cost of losing the dimension of time. It only allows the diagnoses of species in a single place and at a single time. An alternative version of this conception of a species appeals to specific mate recognition systems: species are bounded by mate recognition systems (Paterson 1985). These recognition systems result in genetically isolated populations.

Ecological species

A group of organisms that shares the same adaptive niche (van Valen 1976). The fundamental claim here is that the stability of a species rests primarily on ecological factors rather than on genetic isolation (Ereshefsky 2001, 87). This conception has the advantage of applying equally well to sexual and asexual species, but at the cost of resting on the controversial notion of an ecological niche.

Cohesion species

A group of organisms forming a cohesive lineage (Templeton 1989). The fact that asexual species form bifurcating lineages tells us that there are nonsexual mechanisms causing coherence of discrete lineages (Templeton 1998); mechanisms that sometimes break down in speciation events. This species concept demotes genetic isolation, acknowledging it as just one of the factors that promote the cohesion of lineages.

Phylogenetic and evolutionary species

As with cohesion species, phylogenetic and evolutionary species concepts are agnostic as to particular processes that produce speciation. However, here the diagnostic test is evolutionary rather than ecological. Evolutionary species are lineages of organisms with their "own evolutionary tendencies and historical fate" (Wiley 1978).

Cladistic species

Cladistic definitions identify species with clades of organisms with distinct taxonomic characters (Mishler and Donoghue 1982; Cracraft 1983; Ridley 1989). For example, a monophyletic species is the least inclusive monophyletic group that shares at least one unique characteristic. These definitions have a striking consequence that many find profoundly counterintuitive; any lineage splitting whatever causes the ancestor species to go extinct. The domestic cat becomes extinct if a pair of cats marooned on an island establishes a population with a single distinctive characteristic.

Evolutionary, phylogenetic, and cladistic species concepts tie species-hood to the bifurcation of evolving lineages. But they are deliberately agnostic about the causes of the speciation events that give rise to phylogenetic structure. They stand or fall by the strength of that pattern and by the utility of the cladistic methods that detect it. But while these ideas are neutral on the mechanisms of speciation, they could be tied to further claims about the process or processes that give rise to that structure. However, there seems not to be a single mechanism responsible for lineage bifurcation. Indeed, one natural interpretation of much of the species debate is that it reflects our increasing knowledge of the many mechanisms underlying diversity and differentiation.

John Wilkins has developed a helpful way of thinking about this diversity of mechanisms and the relationship between them: a three-dimensional conceptual space (2007). One dimension represents the role of chance. Sir Ronald Fisher and Sewall Wright famously debated the role of genetic drift and other chance factors in generating the divergence between populations in a sundering lineage. For example, in vicariant models of speciation a widely distributed ancestral population is divided into fragments by geological changes. These fragments then diverge, and chance is important as they wander morphologically away from one another. So if this model is important, chance plays an important role in much speciation. A second dimension concerns the relative role of intrinsic and external factors when selection does drive differentiation. For example, if hybrids between two subpopulations are less fit, then there will be selection of traits that cause like to mate with like. Features of the evolving population itself shape the selective environment. In contrast, on Mayr's peripheral isolate model, selection will drive differentiation due to external environmental differences between the center of the species' range and the periphery. Wilkins's third dimension focuses on the role of gene flow and barriers to that flow. Mayr, famously, argued for the importance of geographic isolation in the evolution of differentiation. But there are many models of speciation that allow speciation to take place without geographic isolation; for example, speciation that involves host switching by parasites, and speciation that involves hybridization or chromosomal reorganization (a mechanism quite common in plants).

We will illustrate these points about the diversity of mechanism through a brief discussion of ecological and biological species concepts. As usual, the picture is complex. Some ecological species concepts are deliberately agnostic as to the details of the processes that give rise to speciation. For example, Alan Templeton's cohesion species concept takes cohesion to be crucial in the production and maintenance of species, but he accepts that there are many biological processes that generate cohesion. Leigh van Valen's ecological species concept ties species to niche occupation. However, the relationship between species and niches is very complex. It was once supposed that communities were organized in ways that made a variety of roles or occupations available to be filled (or not) by suitable organisms (a classic example is Elton 1927). A niche imposes demands on its occupants, and these demands explain similarities within and differences across species. But ecologists no longer think that we can assume that communities, built from different species, nonetheless have a common organization, that, say, a temperate rainforest in British Columbia will make available the same array of occupations as one on New Zealand's west coast. And so niches are now defined by the organisms that occupy them, by vectors of their resource requirements (Griesemer 1992). Moreover, organisms alter both their own physical and biological environment and those of others. Trees stabilize soils; moderate storm impacts; and provide shelter, resources, and concealment for a host of other organisms (Jones et al. 1997; Lewontin 1985; Odling-Smee et al. 2003).

So the niches of some species depend largely on the geology and meteorology of their habitat and the ecological milieu that determines their place in the food webs and nutrient cycles of which they are a part. In other species, selection for "ecological engineering" plays a much stronger role. Earthworms, for example, are adapted to an aquatic environment. They only survive out of water because they have evolved the ability to alter their own environment. They reduce surface litter; aggregate soil particles; and increase levels of organic carbon, nitrogen, and polysaccharides, which enhances plant yields and improves porosity, aeration, and drainage. In so doing they co-opt the soils they inhabit and the tunnels they build to "serve as accessory kidneys and compensate for their poor structural adaptation" (Odling-Smee et al. 2003, 375). Thus niche occupation appears much more active in some species than others. Furthermore, the idea that each species has a unique niche is more plausible in some cases than in others. Scavenging generalists battle it out in a widely contested and "open to all comers" niche. By contrast, figs and fig wasps have coevolved; each provides an essential and specific service to the other.

The idea that a species lives in, and is shaped by, a unique niche turns out to glide over a complex and variable set of relationships between species and environments. The same is true of the apparently straightforward idea of reproductive isolation. It turns out to be a cover-all label for a large variety of prezygotic and postzygotic interactions that largely keep lineages separate. The great strength of the biological species concept is that reproductive isolation and the resulting restriction in gene flow are real and important facts in the evolution of populations and metapopulations. However, there are many ways in which gene flow can be restricted. Australian immigrant species that are wind-borne to New Zealand shores face the blustery Tasman Sea as a barrier to future gene exchange with the Australian populations from whence they came. Here, reproductive isolation is extrinsic and geographic. Similarly, it has long been recognized that meteorological factors are crucial, particularly for the transport of insects (Tomlinson 1973). Compare this allopatric case with sympatric speciation in plants due to polyploidy. Isolation here is neither geographic nor ecological but genetic.

When we focus on the mechanisms that generate and prevent differentiation, and the ways these vary across differing lineages, the "common currency" problem looks pressing, and the case for species richness as a general purpose measure of biodiversity looks to be in trouble. Faunas and floras are compiled using different species definitions, and, as we have pointed out, those inventories are not stable in the face of differing species concepts. We might easily be led to think (for example) that the amphibian fauna of Sri Lanka is richer than that of Sumatra, because the first but not the second has been revised with the use of molecular data, and using a cladistic criterion for species recognition. But problems seem to remain even if we could escape the practical limits imposed by species lists compiled using divergent species concepts. There are competing species definitions, because differing accounts of species and speciation seem to fit differing lineages. Mayr's biological species definition really does seem to work well for his birds.

What if we accepted this, and tried to work with species concepts appropriate to differing lineages? We might decide that the standard biological species concept fits vertebrates; the cohesion concept fits vascular plants; a criterion based on specific mate recognition systems fits arthropods. Being cautious, we might then decide that a vascular plant species in Sumatra is not equivalent to a Sri Lankan fruit fly, but at least it is equivalent to a Sri Lankan vascular plant. Following this strategy, we would then fractionate our measures of species richness: comparing vertebrate richness to vertebrate richness, vascular plant richness to vascular plant richness, arthropod richness to arthropod richness. Even this cautious approach seems problematic. We still have an integration problem; conservation decisions require us to make overall assessments of biodiversity and of the relative biodiversity of habitats. Moreover, as we have just seen, even two good "biological species" might not be units of the same kind if one is isolated merely by extrinsic factors, while the other is genetically incompatible with its sibling species. Despite this line of argument, in the next section we suggest that there is a lot to be said for a species-based approach to biodiversity.

2.3 THE EFFECT OF SPECIATION

Speciation is problematic, as we have just seen, because of the variety of mechanisms through which one lineage can become two. All that these mechanisms have in common is their effect; dividing lineages acquire independent evolutionary trajectories. It is also true that independence is a matter of degree. Moreover, the transitivity of gene flow makes it possible for there to be gene flow between two groups of species even though members of one group are not able to interbreed with members of the other group. In some cases, such as domestic dogs, we have been happy to accept such heterogeneous metapopulations as single species. In other cases (generally when gene flow is a little better behaved) taxonomists have accepted that populations are distinct species that are nonetheless genetically linked metapopulations (an example is described in Box 2.2.)

BOX 2.2: Reproductive Isolation

Vicariance events leading to prolonged geographical isolation are a major cause of speciation. However, when an existing population expands to surround a large uninhabitable region, strange partial speciation events can occur. The standard-bearer for the group of so-called ring species has until recently been the herring gull (*Larus argentatus*) complex, which has a circumpolar distribution in the Northern Hemisphere. However, recent work suggests that this much loved example is not in fact a ring species at all (Liebers et al. 2004). Thankfully, an understudy to the role exists in the form of the Eurasian greenish warbler complex whose range expanded around the margins of the arid Tibetan Plateau. The flanks of the initial population now overlap on the northern edge of the plateau where they differ in both plumage and song, but do not interbreed (Irwin et al. 2005). The western flank is now classified as *Phylloscopus viridanus* and the eastern as *Phylloscopus plumbeitarsus*.

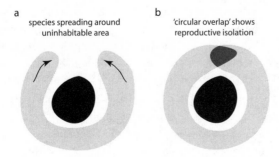

a
species spreading around
uninhabitable area

b
'circular overlap' shows
reproductive isolation

FIGURE 2.1. Speciation by circular overlap. After Helbig (2005).

The fact that speciation is a matter of degree is a potential obstacle for those using species richness as a measure of biodiversity. However, it is also an explanatory windfall in the investigation of speciation and its consequences. For there is a deep connection between speciation and phenotype change. This connection was first pointed out by Douglas Futuyma (Futuyma 1987), though in recent years Niles Eldredge has been chiefly responsible for developing it (Eldredge 1995; 2003). It is central to Eldredge's account of why the life history of most species has the classic "punctuated equilibrium" pattern; the distinctive phenotype of a species evolves as that species comes into existence, with little further net change over that species' lifetime. Phenotypes stay roughly constant over time; that is, variation over time is not significantly greater than variation within a population at a time. Phenotypes are stable despite the fact that local populations do adapt to their specific conditions, sometimes quite rapidly (Thompson 1999, 12). But the metapopulation dynamics of species typically result in local adaptations being lost; they are ephemeral. For species are typically *ecological mosaics*. The common brushtail possum (*Trichosurus vulpecula*) is found in communities as varied as cool temperate New Zealand rainforests, inner Sydney suburban gardens, and eucalypt woodlands. Thus their relations with those organisms on which they feed, those with which they compete, and those that threaten them with predation, all vary importantly from community to community. Their physical environments vary, too. So there is no single set of selective pressures acting on the possum population as a whole. Possums, of course, are exceptionally tolerant and have an unusually broad geographical distribution. But roughly the same is true of species with narrower geographic range. Variation in time and local, small-scale heterogeneity typically expose distinct populations to varying mixes of selective forces.

For the most part, then, species are geographic and ecological mosaics. Species do not have niches. Instead, they are ensembles of populations, each with its own niche. Thus, to the extent that these populations adapt to their circumstances, coming to differ from the original phenotype of the species, they will do so in different ways. However, while these populations are ecologically distinctive, they are not demographically isolated (not, at least, for long periods of time). So there is gene flow between local populations, and between populations on the periphery of the species range and populations in the center of that range. To the extent that local adaptation depends on specific gene combinations, gene flow makes local adaptations vulnerable to dilution effects.[4] These dilution effects will often be decisive. For the gene flow is from populations without these new gene combinations, and these source populations are not under selection to acquire or retain these new genes. Furthermore, the demographic center of gravity usually consists of populations with the original phenotype of the species. Thus, to the extent that an adapting population is well connected demographically to the rest of the species, the local adaptations it acquires are liable to be lost.

That said, demographic connections between metapopulations do not invariably stabilize phenotypes. Some selective impacts will be of the right spatial scale to generate change. Climatic and other changes in the physical environment might well generate coarse-grained selective forces, affecting all or most of the species' populations. Tectonic and other geological forces that alter the basic structure of the landscape (uplift, erosion and deposition, sea level changes) may well exert consistent effects over large areas. So changes that are adaptive throughout the species' range can become established, but as Futuyma notes (1987, 468), these changes are likely to be uncommon. Hence, the connection between phenotype change and speciation.[5] Speciation is not required for phenotype change that adapts a population to specific local conditions, but it is often required to make such changes permanent. The distribution of a species through an ecological mosaic, together with gene flow between the fragments, acts as a brake on evolutionary change.

Stasis is not permanent, however, in part because environmental change has the potential to release the evolutionary brake imposed by metapopulation dynamics. Climate change and other large-scale physical changes can turn species mosaics into patchworks of isolated populations (see Bennett 1997; Eldredge 1995; Vrba 1993 and 1995). Sometimes the effects of environmental change will not be dramatic: the potential space available to a species might shrink a bit, expand a bit, or shift latitude. If physical barriers do not intervene, the species can shift with it. However, stasis breaks down when environments both change (creating new selection pressures) and a species' range fragments, dissolving the metapopulation by chopping it into its component populations. A local, isolated population is not ecologically fragmented. The vast majority of such small populations will go extinct. But if in these fragments the population is not so small that genetic variation is sharply reduced, selection can act, and act without counterbalance, from homogenizing gene flow from neighboring populations. For there are no neighboring populations. While many fragments go extinct, a few will survive as new species. Many species emerge through a life cycle: from population $\Rightarrow$ metapopulation $\Rightarrow$ isolated population $\Rightarrow$ incipient species. This life cycle is itself one important mechanism of evolutionary change. As many phenomenological species are the product of this life cycle, the species category is something like a natural kind. In counting species and comparing the species richness of different regions, we can, roughly, compare like with like.[6]

2.4 SPECIES AND BIODIVERSITY

The living world is organized into phenomenological species: recognizable, reidentifiable clusters of organisms. This fact makes the production of bird and butterfly field guides, identification keys for invertebrates, regional floras, and the like, all possible. In this chapter we have embraced the common biological wisdom that phenomenological species richness captures a crucial dimension of biodiversity. There are many routes through which one population can become demographically isolated from, and hence evolutionarily independent of, populations that were once sources and sinks of its own genes. But the fact of isolation and evolutionary independence is of immense importance to the fate of local adaptation in such populations. So the phenomenological species richness of a region is, in an importance sense, a catalogue both of phenotypic variety and of the potential evolutionary resources available in that region. There is an important difference, on this picture, between a single widespread and phenotypically variable species (like the common brushtail) and a set of closely related species. The available phenotypes, population sizes, and ecological roles might be exactly the same. But one set of phenotypes will be entrenched by speciation mechanisms, and hence will survive minor ecological changes that increase migration rates across the landscape; the other set is much more fragile in the face of relatively minor ecological change. So it does not just matter what phenotypes are present; how they are bundled into species is also important. In effect, we have defended a version of an evolutionary species concept, and we accept that the collection of independently evolving lineages in a region is a key component, perhaps the key component, of that region's biological diversity.

That said, we need to add some important qualifications. The identification of phenomenological species with metapopulations in partial stasis holds good only for some chunks of the tree of life. It may not fit plants. It clearly does not fit microbes. Frederick Cohan has developed an account of bacterial species that identifies those species with the units of bacterial evolution.[7] But this involves abandoning the idea that phenomenological species are typically important units of stasis and change; in Cohan's view, phenomenological bacterial species are amalgams of many evolutionary units (Cohan 2002). Even if we set that aside, phenomenological species do not represent equal amounts of evolutionary information and evolutionary potential. In different lineages, there are enormous differences in species richness and morphological diversity. In the next two chapters, the focus changes to the relationship between species richness and morphological diversity. If species richness is only one albeit crucial component of biodiversity, what do we need to add, and for what explanatory, predictive, and practical projects?

3 *Disparity and Diversity*

As we have just argued, in measuring the diversity of a biota, we start well by counting species. Despite all the complications we discussed in the last chapter, species are objective units in nature. Since species come into existence through biological processes that take time, there must be intermediate cases: populations part way through speciation. But aside from such cases, "good" species are objective and observer-independent. A competent systematicist from Alpha Centauri would recognize the long-beaked and short-beaked echidna as different species, whatever odd perceptual equipment and biological interests might have evolved there. In contrast, such an Alpha Centaurian biologist might not even recognize our higher taxonomic ranks. Genus, family, class, order, phylum are conventional. A genus is a monophyletic group of closely related and phenotypically similar species. But no one supposes that there is an objective answer to the question: how similar must species be, to count as members of the same genus? Likewise, the populations out of which species are composed are ephemeral, with ill-defined boundaries. Populations divide and coalesce repeatedly, often in response to quite minor ecological and climatic events. Moreover, they are usually demographically open; genes move between adjacent populations, and there is no telling how much sideways flow counts as the two populations merging into one. In contrast, then, to a genus of which it is a member and the populations out of which it is formed, the short-beaked echidna (*Tachyglossus aculeatus*) is a genuine, objective element of the biota. So species are atoms of diversity (at least for some purposes). But how should we take into account their similarities and differences in measuring the diversity of a system? For, as we noted in 1.3, the diversity of a system will depend both on the number of distinct elements in it and the extent of their differentiation.

In this chapter and the next, we will consider the idea that tracking species number does not track a second important dimension of biodiversity: phenotypic richness. We will focus on the claim that diversity (= species number) does not track disparity (= variation across phenotypes). A biota can be species rich but not very disparate, if the species composing the biota are rather similar. Arguably, many island faunas are more diverse than disparate, for they often derive from a few founder species, and this constrains the variation that evolves. Despite their different beaks, Darwin's finches really are pretty similar birds. Stephen Jay Gould made this diversity-disparity distinction famous in his 1989 classic *Wonderful Life*, and it has generated ongoing controversy.

The phenotypic spread of a biota is important to its evolutionary and ecological future; a phenotypically richer biota is more apt to be biologically prepared for change of various kinds. It can recruit and amplify existing variation to meet change, rather than having to wait on migration or evolution to create new variation. And the phenotypic spread of a biota is also a signal of its ecological and evolutionary past, of the processes that have been important in its making. In this chapter our main interest will be in the idea that high phenotypic biodiversity is a signal of distinctive mechanisms (we discuss the ecological consequences of phenotypic diversity in chapter 6). But whether we are interested in diversity as a signal or as an input to further change, the spread of phenotype is important. What is not clear is whether we have to count it separately (and if so, how we should count) or whether spread is indexed by species-level diversity. While there is more to the phenotype of an organism than its morphology, we will use morphology as our surrogate for phenotype diversity. For we have temporal depth in morphological information. Behavior and physiology leave less of a paleobiological signal. There is no doubt that phenotypes vary from one another in objective and important ways. There is, however, considerable doubt that there is a single metric or framework into which these variations can be placed; a decent multipurpose measure of the phenotypic spread of a biota. As we shall see, a wide range of biological projects suggest the existence of such a framework. But we are skeptics; there is no overall, objective measure of phenotype difference. If there were, the phenetic program in systematics could be revived.

Gould himself explored this distinction in the context of paleobiology, of long-term trends in biodiversity. There has been considerable debate about the impact of biases in the fossil record on our ability to reliably estimate changes in species number over time (Alroy 2000; Foote and Peters 2001; Bush and Bambach 2004). Even so, there is wide agreement that one trend has been an increase in species number over time, despite severe interruptions by catastrophe.[1] Gould accepts this upward trend in species number, but denies that there is any upward trend in disparity, in particular in animal disparity. He bases this argument on one of the most remarkable and controversial episodes in the history of life, the "Cambrian explosion." In a relatively short period in earth's history, a diverse and recognizable fauna first appeared. These Cambrian fossils are not the oldest multicelled animals. There was an older fauna still, the Ediacaran fauna. But the Ediacaran animals were very strange, and the fossils they left are very hard to interpret. So hard, in fact, that some paleobiologists argue that these Ediacarans were a failed experiment in multicelled life. On this view, they left no descendants (for this idea and responses to it, see Seilacher and Buss 1994; Narbonne 1998; Knoll 2003). In contrast, the Cambrian fauna includes clear ancestors of many of the major invertebrate clades, perhaps even an early chordate. Animal life as we know it had clearly evolved by the mid-Cambrian.

This, too, is uncontroversial. The controversy begins with Gould's further claim that the Cambrian explosion was such a rich burst of radiation that animal disparity peaked in the Cambrian. As Gould saw it in *Wonderful Life*, within evolutionary biology there is a standard conception of the dynamics of diversity over time: the "cone of increasing diversity." This standard conception, Gould argued, was a misconception with two sources. One is simply the failure to attend to the distinction between diversity and disparity. With that distinction unnoticed, the upward trend in species number is taken to be an upward trend in disparity as well. A second is a massive failure to appreciate just how strange and varied the Cambrian fauna was. While it does indeed include recognizable ancestors of the modern great clades of animals, it includes, Gould argued, much else as well. Indeed, *Wonderful Life* was written in response to the rediscovery of the richness and strangeness of one slice of Cambrian life, preserved in the extraordinary fossils of the Burgess Shale.

As Gould represents the state of play, to the extent that the standard conception of evolutionary history distinguished between disparity and diversity at all, that standard conception was committed to the idea that disparity trends upward much as species numbers do, a conception he depicted as shown in fig. 3.1. He thinks the reality is very different. For most of the history of life, there was remarkably little change in disparity. Bacteria evolved, and from them eukaryotic cells and even some colonial eukaryote complexes. The plantlike Ediacaran biota and the small shelly fauna of the earliest Cambrian likewise eventually

FIGURE 3.1. The cone of increasing disparity. After Gould (1989, 46).

appeared on the scene. Then, in the early to mid-Cambrian, there was a massive explosion in disparity. Suddenly we see a bewildering variety of large complex animals. There are arthropods, but strange ones indeed compared to the living insects, spiders, and crustaceans, or even compared to the extinct trilobites. Many others are representatives of wholly extinct phyla. For soon afterward much of this rich and strange fauna disappears. As a result, what we see today is a mere remnant of the great disparity that existed in the mid-Cambrian. Despite hundreds of millions of years of speciation and adaptation, mass extinction and adaptive radiation, the biosphere has never recovered that astounding disparity that flowered so briefly, so long ago. The right representation of disparity's history is early radiation followed by decimation (fig. 3.2).

In the next section, we develop Gould's take on the Cambrian and its significance in more detail, explain the importance of the issues Gould raises, and outline some of the challenges to his view. In 3.3 we sketch recent developments in understanding the Cambrian, and their implications for the Gouldian picture. In the next chapter, we focus on the most central challenge of all: whether disparity is a genuine dimension of biodiversity.

3.2 HOW DISPARATE WAS THE CAMBRIAN FAUNA?

Gould makes two critical claims about the Burgess Shale fauna of the Cambrian. First, he claims that those invertebrates were more disparate

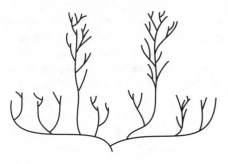

FIGURE 3.2. Decimation and diversification. After Gould (1989, 46).

by far than contemporary animal life. Diversity peaked then, first because the middle Cambrian saw a massive radiation of animals, and second because the disparity lost through extinctions since the Cambrian has not been replaced. Evolutionary mechanisms still generate new species and new adaptive complexes. But they do not generate new ways of organizing the overall body plan of animals. The Cambrian seas, Gould tells us, contained a wide variety of creatures whose fundamental body architecture was profoundly different from those of extant organisms. Moreover, we will never see the likes of Cambrian disparity again, for evolutionary mechanisms are no longer capable of producing fundamentally new body architectures. This is why Gould depicts the history of life as diversification followed by decimation rather than one of diversification, decimation in mass extinctions, followed by the evolution of further diversity: a history as depicted, for example, in figure 3.3.

The Burgess Shale fauna was originally described by Charles Walcott, who "shoehorned" (as Gould put it) a remarkable set of fossils into known taxonomic groups. What he saw in the Burgess fossils was evolution as usual. There the story might have ended but for the work of Harry Whittington, Derek Briggs, and Simon Conway Morris, who in the 1970s undertook the arduous task of reexamining the Burgess fossils (chronicled in Conway Morris 1998). Their findings were remarkable. According to them, many of the Burgess organisms did not belong to known phyla. Moreover, the Burgess arthropods were also particularly problematic. Until the reexamination of the Burgess specimens it was assumed that all of the millions of living arthropod species, as well as the many millions of arthropods that are now extinct, belonged to four great groups ("classes"): insects and their relatives; spiders and spiderlike arthropods; crabs, lobsters, and their close relatives; and the long-extinct trilobites.

Many of the Burgess arthropods did not belong to any of these four great groups. Moreover, it also seemed clear that they could not be shoe-

FIGURE 3.3. Decimation and subsequent replacement. Note that this version of evolutionary history shows lost disparity being replaced by subsequent evolution. The base of this figure is the same as Gould's decimation and diversification graphic (Gould 1989, 46), but on the two surviving lineages extant clades replace the missing diversity.

FIGURE 3.4. *Anomalocaris.*
By permission of Sean Carroll.
Drawn by Leanne Olds.

horned into a single extra taxon of the same rank. The morphologies represented in the Burgess Shale are simply too disparate for them all to be aggregated into a single extra arthropod group. To those familiar only with modern arthropods, the giant *Anomalocaris* (see fig. 3.4) with its massive frontal grappling hooks and a circular mouth made up of overlapping plates with sharp prongs extending into the oral cavity, appears like a creature from another planet. Even more strange is the five-eyed *Opabinia regalis* (see fig. 3.5), looking like a prehistoric vacuum cleaner, with its frontal feeding nozzle tipped by a single claw.

In Gould's view (and in the initial view of those responsible for the re-description of these fossils) some of these "weird wonders" represented whole new phyla. Yet, as far as we can tell, very few phyla have evolved since the Cambrian.[2] So one way to read the Cambrian fossil record is that it shows that many phyla of metazoans evolved in the Cambrian, and perhaps none at all have evolved since the Cambrian. We still see the evolution of new adaptive complexes, new adaptive solutions to environmental problems. But we do not see new body plans to house those adaptations. The Cambrian was a unique radiation, a radiation in fundamental organization. Inevitably many of those remarkable early organizations did not survive. The descendants of *Anomalocaris*,

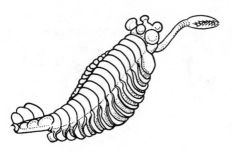

FIGURE 3.5. *Opabinia regalis.*
By permission of Sean Carroll.
Drawn by Leanne Olds.

Opabinia, and other weird wonders are not with us today. Nor are there extant creatures like them. The reason for this, says Gould, is that the body plans that exist today have become fixed—entrenched. Modern evolution is a variation upon a relatively small number of themes.

Gould's claim is clearly important. A good theory of evolutionary mechanisms must explain the patterns we find in the history of life, and Gould's scenario challenges our standard picture of those mechanisms in two ways. His view of Cambrian disparity sharpens the problem of giving an explanation of the Cambrian explosion itself. The more rapid we see that explosion to have been, and the greater the disparity we take it to have generated, the more challenging this explanatory problem becomes. The biodiversity of the Cambrian is a signal of the evolutionary mechanisms that built the Cambrian biota. Gould and his allies (for example, McMenamin and McMenamin 1990) argue that one aspect of this biodiversity—its rapidly evolving and extensive disparity—is a signal of the operation of very unusual evolutionary mechanisms. Both the speed of change and the phenotypic variation that result are important in this argument. We will see that both strands of the case have been challenged.

Gould's picture of the Cambrian poses a second explanatory challenge. In his view, there was a contraction in the evolutionary possibilities open to those lineages that survived the end-Cambrian extinctions: they were no longer potential ancestors to a Cambrian-style radiation. The Cambrian radiation is essentially a radiation of bilaterian animals, animals with a front-to-back axis of symmetry and a third layer of cells in the developing embryo. There was no spectacular flowering of the even more basal animals, the sponges and the cnidarians. But in Gould's picture, the Mother of all Bilaterians was ancestor to a wonderfully rich and bizarre family of creatures. There were, and are, many bilaterians in the post-Cambrian seas, but none of them was a potential mother to such a radiation. If his views are right, we need to explain not just the

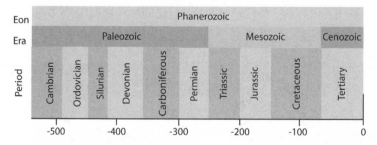

FIGURE 3.6. The geological time scale from the Precambrian to the present day (in millions of years).

rich burst of novelty, but also why that burst cannot be repeated. We return to the issue of constraints on the supply of variation in chapter 5. In this chapter, our focus is on disparity.

If Gould's views about the early history of animal evolution are true, they are important and challenging. However, it is far from clear that Gould's claims are true. One line of argument threatens the whole conceptual foundation of Gould's picture by arguing that there is no coherent, objective notion of disparity. That crucial issue is the concern of the next chapter. A second line of argument does not dismiss the notion of disparity outright. But it threatens the idea that disparity peaked in the Cambrian; this is dismissed as a taxonomic illusion. Mark Ridley, Richard Dawkins, and Dan Dennett have all developed this line in their immediate responses to the argument of *Wonderful Life*. They argued that Gould appealed to inappropriate criteria in arguing that (for example) many Burgess arthropods were wonderfully different from contemporary survivors. Contemporary arthropod systematicists use limb attachment structure and the patterns in which body elements fuse together (tagmosis) to diagnose the great arthropod clades. As Ridley and Dawkins point out, this makes sense only because of the historical contingencies that have made certain morphological characteristics (like spiders' leg number) reliable and stable markers of membership of these great evolutionary families. It is not because (for example) leg number is especially important. Indeed, leg number varies in other clades. Despite their name, centipedes are not all equipped with one hundred legs. We cannot project the significance of these badges of family membership back in time and treat them as surrogates for disparity (Ridley 1990, 1993; Dawkins 2003).

Gould saw this objection coming and tried to head it off (see 1991, especially 414; 1989, 209). Thus in *Wonderful Life*, he notes the problem and then argues:

> If you wish to reject tagmosis as too retrospective then what other criterion will suggest less disparity in the Burgess? We use basic anatomical designs, not ecological diversification, as our criterion of higher-level classification (bats and whales are both mammals). Nearly every Burgess genus represents a design unto itself by any anatomical criterion. (209n)

We think this argument is inconclusive. Not all the weird wonders of the Burgess redescriptions have stood the test of time. *Hallucigenia*, famously, turned out to have been reconstructed upside down. The right way up, it is probably a velvet worm—a representative of a living phylum (Hou and Bergstrom 1995). Very recently, another creature of mysterious shape and affinities, *Odontogriphus*, seems to have found a home with the mollusks, thanks to the discovery of more specimens. But many do look genuinely, deeply strange. That may just be the pull of the recent: old fossils, and especially those from evolutionary lineages without living descendants, will look strange. As we shall see in the next section on contemporary work on the Cambrian explosion, the argument about taxonomic illusion has been reformulated in a new and powerful way. So we discuss the contemporary view of the explosion and its implications for the diversity-disparity distinction in the next section.

3.3 FOSSILS IN A MOLECULAR WORLD

We now have a much-improved fossil record of Cambrian and Precambrian life. There are now exceptional sites in China and Greenland recording Cambrian fauna, together with some remarkably detailed microfossils, including fossils of metazoan embryos from the late Proterozoic.[3] While there are inevitably limitations (reviewed in Butterfield 2003), even in the data provided by these exceptional sites, our picture of the Cambrian metazoan radiation is more complete, and so is our picture of its relation to the Ediacaran fauna that preceded it. It is now clear that the main radiation of fossilized Cambrian fauna took place considerably after the base of the Cambrian, and did not coincide with the eclipse of the Ediacarans. Nor did it take place in the blink of an eye. At the base of the Cambrian, trace fossils change character, hinting at some kind of morphological innovation. The spectacular fossils of the Burgess Shale are almost 40 million years younger than these traces.

The boundary between the lower Cambrian and the Ediacaran (the late Neoproterozoic) is now reckoned to be around 542 million years ago (Whitfield 2004). The Cambrian explosion is a complex phenomenon (see fig. 3.7). It begins in the so-called Tommotian stage, which began about 530 million years before present (Valentine et al. 1994; Babcock et al. 2001). Insofar as we can tell from the fossil record (and this is an important qualification) the history of animal diversification looks something like this: from 600 million years ago to around 550 million years ago there is evidence of metazoan life: metazoan embryos, rare body fossils, and a few trace fossils. But it is sparse. From then to the base of the Cambrian there are quite rich Ediacaran assemblages with some evidence of an important behavioral transition around the base of the Cambrian. In the early Cambrian (in the Manykaian stage) small shelly fossils increase in diversity, but it is not until the

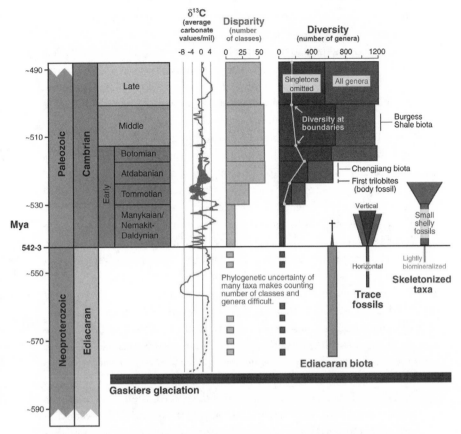

FIGURE 3.7. Complex anatomy of the Cambrian explosion. Dates from Grotzinger et al. (1995), Landing et al. (1998), Gradstein et al. (2004), and Condon et al. (2005). Neoproterozoic carbonate carbon isotope curve from Condon et al. (2005), early Cambrian curve largely from Maloof et al. (2005) but also from Kirschvink and Raub (2003), and middle and late Cambrian from Montañez et al. (2000). Disparity from Bowring et al. (1993). Diversity based on tabulation by Foote (2003) derived from Sepkoski's compendium of marine genera (Sepkoski 1997, 2002). From Marshall (2006). Reprinted, with permission, from the *Annual Review of Earth and Planetary Sciences* 34 (2006), www.annualreviews.org.

Tommotian that the explosion becomes obvious, with the first mollusk and brachiopod body fossils. The spectacular fossils of Chengjiang date at around 520 million years ago and the Burgess fauna is up to 15 million years younger (Knoll 2003; Valentine 2004, 170). The Ediacaran fauna may well have been extirpated by the evolution of bilaterian predation (as McMenamin and McMenamin [1990] have conjectured), but this hypothesis gets no support from Ediacaran extinctions coinciding with a metazoan radiation.

So the rocks seem to tell us that the main animal radiation took place in the early to mid-Cambrian. The molecules seem to tell a different story. Molecular clocks work by identifying homologous regions of the genome across different metazoan lineages (for an introduction, see Delsuc et al. 2005). We then establish the extent to which these regions have diverged from one another and calibrate the rate at which these genomic regions change by comparing lineages whose divergence is well dated (for example, splits among the main vertebrate lineages).[4] The striking outcome of molecular clock dates of divergence has been to push metazoan lineage divergence deeper in time, although there is not yet consensus as to the date. A range of different genomic regions has been used in cross-lineage comparison and calibration, and these give different dates. Some are well back into the Proterozoic (Wray et al. 1996; Hedges 2002). Kevin Peterson and his colleagues argue that the deep dates depend on using the slowly evolving vertebrates for calibration (Peterson et al. 2004). However, even calibrating using faster evolving lineages, the sponges and cnidarians branch from the metazoan lineage significantly before the Cambrian. "Relaxed clock" methods, which do not assume that sequences diverge at the same rate in all the taxa under consideration, also result in molecular clock dates closer to rock dates (Douzery et al. 2004), but these have their own problems (Bromham 2006). Recent work has reduced the mismatch between molecular and fossil data (Jermiin et al. 2005; Rokas et al. 2005); nonetheless, there still is a mismatch.

We noted in 3.1 that the case for thinking that unusual evolutionary mechanisms acted in the early Cambrian depended both on the rapidity of the Cambrian radiation and on the richness of Cambrian disparity. In the light of both fossils and molecules, the Cambrian radiation now seems less rapid than Gould had supposed. Recent developments suggest that the Cambrian may have also been less disparate. Gould used higher taxonomic categories as a surrogate for disparity; he argued that if we classified Burgess Shale arthropods using the principles we now use when confronted with new, living arthropods, we would recognize new arthropod classes and even new phyla. A new integrated theory of the metazoan life suggests how to place the problematic Cambrian fossils into the tree of life, and it explains why they seem strange without actually being strange. If this line of thought is right, morphological diversity in the Cambrian has closely tracked Cambrian phylogeny. Gould's conclusions about the existence of new classes and phyla largely reflect our temporal distance from those ancient organisms. But this reply depends on our being able to place our strange-seeming Cambrian creatures in the tree of animal life, so to that we now turn.

At the time of *Wonderful Life*, the order of branching of the metazoan phyla was extremely controversial, in part because their body plans were so distinct. There was an orthodox view of the history of the Metazoa, but it was coarse-grained, and many relationships among the phyla were unresolved or controversial. The fundamental split was between the basal, embyrologically simple (developing from just two germ layers) sponges, cnidarians, and ctenophores in one group and the bilaterians in the other. The bilaterians are morphologically and embryologically more complex; they have three cell layers in their embryo—they are tripoblasts. They have front-to-back and up-to-down axes of symmetry; they have an organized nerve system with a frontal concentration of nerve cells; most of them have a through-gut with a distinct mouth and anus; most of them have a true, muscle-walled body cavity. And (it turned out) they have an enlarged set of Hox genes. The traditional view saw these morphological innovations as evolving by stages. For not all bilaterians have a true body cavity. The platyhelminthes—the flat-worms—do not. Moreover, there is a cluster of poorly known pseudo-coelomate phyla whose body cavity is partial, and whose development is embryologically distinct from that of the true coelomates. So on this classical view, these groups were evolutionary way stations on the road to the full-deal bilaterians, which in turn fissured into two main groups distinguished embryologically: the deuterostomes and protostomes.

The affinities of some phyla remain unclear, but molecular systematics has led to a consensus about the coarse-grained pattern of metazoan evolution, and on how to fit much of the eccentric Cambrian fauna into this picture. This new phylogeny casts doubt on the traditional view of the stable and gradual development of metazoan body plans.

The picture of the base of the metazoan radiation has not changed. However, within the bilaterian animals, molecular systematics has led to a revolutionary change. On previous views of animal evolutionary history, the evolution of the true body cavity—the "coelom"—was seen as progressive and gradual. The animal phyla lacking a true coelom (the acoelomate flatworms, the platyhelminthes) were seen as primitive, resembling the Mother of All Bilaterians in this respect, and those phyla with partial cavities likewise were seen as more primitive than the true ceolomates. Molecular data have not supported this view. The pseudo-coelomates are not a clade; they are secondarily simplified, as are some of the flatworms. They are not a clade either; some of the worms without body cavities are indeed primitively simple and are the most ancient surviving split within the bilaterians, a sister group to all the rest.[5] But others are secondarily simplified. Neither they nor the pseudo-coelomates are way stations on the road to the full bilaterian innovations. They are lineages that evolved a coelomate body plan but then lost it again.

The traditional deuterostome/protostome split is supported. In protostome development, the mouth is the first opening in the embryo; examples are mollusks, annelids, and arthropods. In deuterostome development, the first opening is the anus. The chordates, echinoderms, and hemichordates are deuterostomes, and so this group does form a clade, though it remains uncertain when they branched from the other bilaterians. Moreover, some phyla previously considered deuterostomes have been exiled to the protostomes. The picture of protostome evolutionary history now divides into two nontraditional branches. One is the ecdysozoans, linked together morphologically mainly by the structure of their cuticle (but by many molecular characteristics). The other is the Lophotrochozoa, a clade consisting of those phyla with a lophophore, the distinctive feeding apparatus of the brachiopods and phoronids, together with the phyla of animals with trochophore larvae—ciliated, mostly planktonic larvae. (For a good summary of this recent consensus, see Eermisse and Peterson 2004). While the relationships within these major groupings and the absolute timing of branch points remain matters of great uncertainty, the overall topography of the metazoan clade is now reasonably clear.

This changed and much better confirmed view of metazoan history is of importance to the disparity debate in two respects. First, it undermines the idea that a body plan is a single, integrated morphological-developmental system. Whittington and his co-workers tend to represent the Cambrian lineages as a set of lineages without well-defined origins or clear affinities.[6] Such depictions emphasize the differences between the phyla and suggest that phylum body plans are package deals rather than suites of traits that happen to co-occur. Thus Lindell Bromham takes one hypothesis about the Cambrian to be the idea that phyla are not conventional. Like species they are real; they "represent a fundamental level of organization" (Bromham 2003, 148). This view, she suggests, supposes that body plans are evolutionarily fixed, rather than subject to assembly then reassembly. Once evolved they are a package, resistant to further evolutionary transformation.[7] Bromham argues that DNA-based phylogenies disconfirm this view of phyla by showing, for example, that the basic bilaterian body plan once evolved is not fused. Echinoderms evolved from ancestors with a bilaterian body plan, yet they are radially symmetrical. Likewise, the evolutionary simplifications of some flatworms and the pseudo-coelomate lineages show that the bilaterian body plan is not fixed once in place. We will see in chapter 5 that there might be something to the idea that body plans, once they evolve, are stabilized and become difficult to change. This idea is an important theme of Bill Wimsatt's work (see Wimsatt 2007). But this idea of stabilization does not show there is anything special about phyla, about, say, the arthropod rather than the trilobite body plan. The body organizations we take to be distinctive of the metazoan phyla are not especially, uniquely stabilized. A phylum is a large monophyletic chunk of the tree of animal life, and the organisms in a phylum will resemble one another in various ways due to their shared deep ancestry. To be told that a biota includes representatives from, say, the arthropods, mollusks, and bivalves is to be given useful information (in contrast, say, to being told that it contains organisms used in Wiccan spells). But phyla are not objectively countable units. After all, the idea of a body plan is fundamentally hierarchical. Cephalopods are mollusks. There is a cephalopod body plan, and a mollusk body plan, and the first is a version of the second. But there is nothing objective that determines which of these organizations, if either, characterizes a phylum.

So we should be cautious about inferring phenotypic disparity from traditional taxonomy. We should be especially cautious if the animals are ancient. This new phylogeny shows how the Cambrian fauna can be integrated within the tree of life, and this integration predicts that the Cambrian fauna would seem to be very disparate, even if it were not. The crucial distinction is between the stem and the crown members of a lineage. This distinction is best explained through an example, and we will borrow Andrew Knoll's example of the divergence between the arthropods and their (possible) sister phylum, nematodes (Knoll 2003, 187–90). These are both members of the Ecdysozoa, so they share the molting cuticle characteristic of that clade as well as its genetic and developmental signatures, but apparently not much else. Arthropods are segmented, with jointed appendages and an external skeleton made from chitin. Nematodes are a species-rich clade of tiny worms, tapered at both ends. The joint ancestor of this clade—their last common ancestor—would have resembled neither. It would not have possessed the distinctive suite of arthropod characteristics, but neither would it have had the radically simplified anatomy (compared to many bilateria) of the nematodes. So consider the evolutionary history of the arthropod lineage leading from this last common ancestor to crustaceans, insects, and spiders. On this lineage the distinctive characteristics that unite the arthropods—segmentation, chitinous skeleton, jointed appendages— would have evolved. But not all at once. Perhaps the order was chitin, then segmented body, then jointed appendages.

It follows that this tree will have two important features. First, there will be *stem group arthropods*: arthropods in the lineage that lie between the last common ancestor of the nematode-arthropod clade and the first arthropod with the distinctive, defining arthropod characteristics. That arthropod species plus all of its descendants constitute the *crown group arthropods*. The identification of nematodes as the arthropod sister lineage, plus fairly elementary evolutionary considerations, shows that there must be taxonomically aberrant arthropods: species that are members of the monophyletic taxon that began with the nematode-arthropod split but which do not have all the distinctive elements of arthropod morphology (fig. 3.8).

Nematodes, which are so unlike arthropods, make this very vivid. But the distinction between stem and crown arthropods does not depend on this specific phylogenetic hypothesis. The sister clade of the arthropods, whatever it turns out to be, must be morphologically very distinct from the arthropods, for these are such a distinctive group of animals. Hence the last common ancestor of the arthropod/arthropod-sister clade will be unlike "classic" arthropods. Early members of the arthropod lineage—species that lived just after the split—must be taxonomically aberrant stem arthropods, whatever the arthropod sister group may be. For example, on some views the tardigrades are the arthropod sister group (Eermisse and Peterson 2004). These too are very unlike arthropods. They are tiny (no more than 1 mm long) with an ill-defined head and four pairs of leglike appendages. However, these are not jointed in the arthropod manner, but are more like those

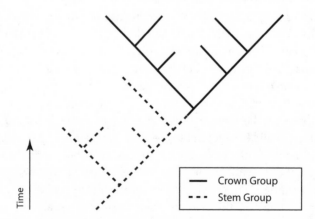

FIGURE 3.8. A phylogeny showing the crown and stem groups. The junction between the two groups marks the evolution of the last taxonomic character that we now take to be diagnostic of the clade. The evolution of other such characters will be scattered through that portion of the stem that is ancestral to the crown.

of velvet worms. And while they have a chitinous covering, it is not a hard exoskeleton (Valentine 2004). So if they are the true arthropod sister, the common arthropod-tardigrade ancestor was not much like any crown group arthropod (fig. 3.9).

Given that very early arthropods were so different from crown group arthropods, there will be a significant evolutionary history between the first stem arthropod and the first crown arthropod. Moreover, that history will itself be a bush: there will be many lineages branching off from the stem-to-crown lineage. All those side branches eventually became extinct, for we would not think of them as side branches had they not. But in some cases, considerable morphological evolution preceded that extinction. It is likely that *Opabinia* and the *Anomalocaris* species represent lineages that split from the stem arthropods, and that saw a good deal of evolutionary change before their extinction. So evolutionary differentiation among stem arthropods—a differentiation we should expect given the great evolutionary distance between early stem and first crown arthropods—generates even more stem arthropods and even more taxonomically unusual ones, as they diverge from the lineage taking us from the Mother of All Arthropods to the crown group arthropods. Crown groups, of course, are lineages whose evolutionary importance is visible only in hindsight when the side branches have

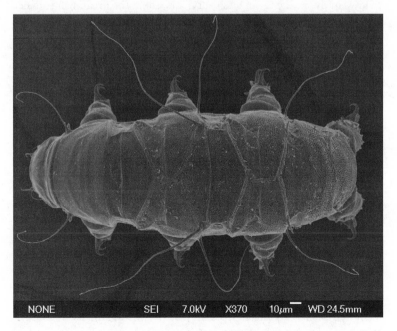

FIGURE 3.9. Tardigrade (*Echiniscus trisetosus*). Photograph by Lukasz Michalczyk and Lukasz Kaczmarek (www.tardigrada.net).

become side branches by extinction. If the lineage represented by *Opabinia* were still living, we would have a different and more inclusive conception of a standard arthropod.

Graham Budd (2003) argues that the hypothesis of special Cambrian disparity is a taxonomic illusion. There is no reason to believe that Cambrian evolutionary dynamics were special in any way. Stem group Cambrian arthropods are bound to have unique characteristics possessed by no living taxa. Yet they will also have some but not all of the diagnostic features of living taxa. This is just the combination that Gould appeals to in his case for Cambrian disparity. An analogous argument can be constructed for the other Cambrian weirdos. Moreover, the Cambrian fossils are old. Budd points out that, as a matter of chance, early branches of a clade are likely to go extinct from time to time. This is the mechanism through which the morphological distance between sister groups increases. These early branches are morphologically intermediate between sister groups, more closely resembling the last common ancestor of a group and its sister. Hence the extinction of early branches increases the morphological distance between the stem and crown members of any group. All else being equal, the longer branches from a common ancestor survive, the less they look like their common ancestor, and the less they look like any randomly selected fellow survivor. The fewer such survivors, then, the greater the chance that each will be quite different from the others, and all will be different from early taxa. These processes imply that the characteristics of the crown group—its distinctive set of shared derived features—will be defined for a highly derived lineage high in the tree. Thus Budd argues, "Given the hypothesis that the base on an extant phylum will be eroded through time, it is clear that the older the fossil is, the more likely it is to fall outside phylum-level classification . . . the pattern demonstrated by the Cambrian fauna therefore seems to be explicable by recourse to the stem-crown group division, rather than to any particular evolutionary mechanism" (Budd 2003, 159; see also Budd and Jensen 2000).

This argument seems to us to show that the taxonomic awkwardness of the Burgess fossils, and especially that of the Burgess arthropods, really is no argument for exceptional Cambrian disparity. On the basis of quite conservative assumptions about the evolutionary processes affecting Cambrian fauna, we would expect to find taxonomically awkward, hard-to-classify fossils. That is especially true to the extent that high-level taxonomy—the way we define phyla and classes—is determined mostly by the morphology of extant organisms. If Budd is right, perhaps phenotypic diversity (supposing it has an independent measure) is tracked by phylogenetic structure. Keeping track of species

richness together with species genealogy would keep track of disparity for free. But phenotype disparity is *not* well-captured by traditional
higher taxa, as their presentist bias makes old taxa look stranger than
they are.

Budd's argument is powerful. But it is also important to be clear
about what it does not show: the fundamental distinction between
crown and stem taxa is neutral on the rate of evolutionary change, and
neutrality cuts both ways. Our ability to explain the systematics of the
metazoan radiation using that distinction is compatible with a highly
disparate Cambrian fauna. We could and should make the distinction
between stem and crown arthropods, even if stem arthropods are as
disparate as Gould, and Mark and Dianna McMenamin suppose. The
stem/crown distinction makes no special assumptions about the nature
of Cambrian evolution. As we have just noted, it is compatible with
conservative assumptions about Cambrian differentiation. But so long
as the Metazoa are a monophyletic clade, it is equally compatible with
the idea that this differentiation was unique. Even if the Burgess fauna
were as rich and weird as Gould suggests, the strange and problematic
Cambrian fossils would still be members of stem groups of extant metazoan lineages: they are bilaterian branches.

The Budd-Jensen argument suggests that, in thinking about the
disparity of animal life, we need to guard against taxonomic illusion:
over-estimating early disparity because early fossils are hard to fit into
taxonomic schemes designed to fit extant organisms. If the Cambrian
fauna were indeed highly disparate, we could still construct a well-
confirmed phylogeny with a stem/crown distinction, one showing (for
example) where *Opabinia* and *Anomalocaris* fit into the stem arthropods.
Equally, we can construct a phylogeny if Cambrian disparity slowly
increases over time. But is there any reason *independent of taxonomic
awkwardness* to suppose that Cambrian fauna were unusually disparate?
This takes us to the problem of morphospace: the idea that we can represent morphology as a multidimensional space, with each dimension
of that space corresponding to a variable morphological feature. If there
is a space of animal form, the disparity of life at a time is the volume of
that space occupied by life at that time. There would then be an open
question about the covariation between species richness and the
occupation of morphospace. While conceding that we are yet to develop an explicit characterization of morphospace, Gould suggests that
the Burgess arthropods are highly disparate in just this sense (Gould
1991). In the next chapter we explore the idea of defining phenotype
biodiversity via the occupation of morphospace, and the connection
between morphospace and species richness.

4 *Morphology and Morphological Diversity*

We have been considering the relationship between species richness and phenotype diversity, in Stephen Jay Gould's useful terminology, between diversity and disparity. Diversity depends on, and hence is a sign of, ecological and evolutionary mechanisms (speciation, migration, and local extinction all influence the diversity of a regional biota). That diversity, once in place, then constrains future change. As we shall see in chapter 6, the diversity of local systems is intimately tied to regional species richness. These processes also build the disparity of regional and global biotas. Phenotypes change through local adaptation, migration, and adaptive plasticity. That disparity, once in place, constrains further change, as new phenotypes allow organisms individually and collectively to shape their environments in new ways. So diversity, disparity, and the relationships between them matter.

In the last chapter, though, we saw that despite the intuitive plausibility of this distinction, it is difficult to make the notion of disparity empirically and theoretically tractable. A central theme of this chapter is that, while species richness does not determine morphological disparity, disparity is conceptually tied to diversity. Patterns in speciation anchor the features of phenotypes we can meaningfully measure and compare. The discussion here will be an echo of that in 1.2, where we discussed the phenetics program in taxonomy. We argued there that similarity and difference must be defined with respect to particular characteristics or traits. That same issue arises for disparity, and we claim that it can be solved only by relativizing disparity to particular clades.

One main message of chapter 3 was that phenotypic biodiversity in general, and morphological biodiversity in particular, is not well captured by Linnaean taxonomy. The initial metazoan radiation is crowded with stem group organisms. Predictably, these appear bizarre to us because stem group organisms *just are* aberrant early forms of modern taxa. The Cambrian forms may well have been uniquely disparate. But we cannot use traditional taxonomy—counting the orders, classes, and phyla present in some biota—to capture the morphological disparity of that biota. The Cambrian example makes the problem of capturing disparity vivid. But the point it illustrates about the limits of the Linnaean system for capturing phenotype variation is general. Unlike species, higher taxa (genera, families, orders, classes, and phyla) are not objective features of the natural world.

How might we do better? In his 1991 article for *Paleobiology*, Gould suggests an alternative approach: we should capture the extent of, and changes in, morphological disparity as changes in the occupation of a "morphospace." A morphospace represents the disparity of a biota by defining a dimension for each morphological characteristic of the organisms in that biota. If there are, say, three characteristics that matter (as there were in a famous application of this idea to shell morphology), then we would get a three-dimensional space. The actual trait values would then determine how much of that space was actually occupied by the biota under consideration.

Spatial metaphors have often appealed to those thinking about biodiversities of various kinds. Richard Dawkins's "genetic space" (1986, 73) and Daniel Dennett's Library of Mendel (1995) were attempts to represent phenotypic and genetic possibility. But these are thought experiments, as George McGhee (1999) noted. They are conceptual models, probing the scope and limits of evolution, rather than attempts at modeling actual biological organisms or formulating testable empirical hypotheses about particular biological systems. In this chapter, we will consider attempts to put spatial representations of morphological diversity to real empirical work. We think this (still young) tradition is impressive. In this chapter we explain why, but we also discuss the limits on these representations of phenotypic diversity.

4.2 MORPHOLOGICAL DIVERSITY

The simplest and most direct approach to phenotype disparity is simply to measure the traits of interest and compare the results. If we were interested in the variation in length in two clades of fish, it would be perverse to count the species in each clade. We need a phylogeny to know which fish belong to each clade, but once we know that we should sample each clade, measure the fish, and analyze the raw length data.[1] Such simple measures of length extent allow us to make genuine morphological comparisons. However, some scientific purposes require more sophisticated metrics of morphology. What to measure and how to compare these measurements become more pressing and difficult questions. In sciences such as paleobiology and developmental biology our main concern is often not with changes in simple distances and volumes but rather with changes in shape.[2] These changes signal new adaptations and migration between niches. For example, one of the great puzzles of vertebrate paleobiology was to understand the path through which fins changed to limbs; a crucial change in mediating the vertebrate invasion of the land.

One promising approach to representing change in shape is called "landmark analysis." Landmark analysis centers on the identification of homologous structures in different species or in the same species over time (Bookstein 1991). Thus although mammals differ widely from one another, they share many homologous structures that can be mapped in a way that allows us to quantify the amount of deformation of their shapes that would turn one into another (see, for example, fig. 4.1).

Of course, the more distantly related that organisms are, the fewer homologies they share. Thus landmark analysis faces severe limitations

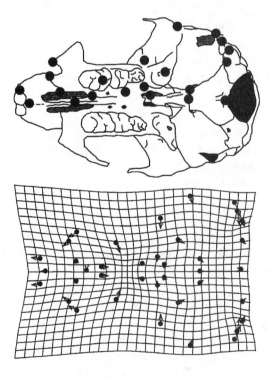

FIGURE 4.1. Landmarks sampled on a skull (above). The grid shows the interpolation of changes between landmarks based on the analysis of displacements between different samples. This represents only the movement of the landmarks with the subsequent distortion of the overall form. From Zelditch et al. (2004). Copyright 2004, with permission from Elsevier.

in comparisons, for example, between protostomes (such as *Opabinia*) and deuterostomes (such as human beings). Furthermore, reliance on homology makes landmark analysis inherently historical. This makes it a powerful tool for the analysis of change within a taxon over time (for example, Eble 2000; Wagner 1995b; Vizcaíno and De Iuliis 2003) as well as between geographically isolated populations of the same species (for example, Hoffmann and Shirriffs 2002). Tools of this kind are quite powerful, as Dan McShea's work on the evolution of complexity shows. There is a consensus in evolutionary biology that the complexity of life has increased over time, though there is no consensus on the nature or cause of that trend. McShea has made this question empirically tractable, by relativizing it to clades, and by developing an operational measure of complexity for homologous organ systems inherited throughout a clade. The vertebrate system, for example, becomes more complex on McShea's operationalization if the number of individual parts (vertebrae) increase, or if the parts become more differentiated, one from another (McShea 1992; 1996). By tracing changes in such systems through a lineage, and by developing these measures of complexity, McShea can then ask whether, on average, organizational complexity in the skeletal systems of vertebrates tends to increase over time. If in most clades and most major organizational systems, complexity tends to increase, we could then conclude that there really was an overall trend to increasing complexity.

However, the same historical foundations render landmark analysis less useful in a comparison of, for example, wing structure between bees, birds, bats, and pterodactyls; their wings are not homologous, each having evolved from a different morphological precursor (see table 4.1). But perhaps the greatest drawback for those who would map large-scale changes in disparity is the fact that the study of landmark data is bound by actual form. It is necessarily derived from existing morphology interpreted in light of known phylogeny. But shouldn't we be able to represent possible as well as actual morphologies? It seems important to be able to represent morphologies we might have, but do not, if we are to identify "gaps in nature." Such gaps are organ-

TABLE 4.1: Wing Precursors in Four Distantly Related Taxa.

Taxon	Wings evolved from
Insects	Gill branches
Birds	Arms
Bats	Hands
Pterodactyls	A single elongated digit

isms whose absence is puzzling. There are behavioral and physiological examples of such gaps: it is prima facie puzzling that there are no eusocial spiders, viviparous turtles, or animals that feed on the aerial plankton. We need to represent possible as well as actual morphological biodiversity if we are to notice forms whose absence should puzzle us. Should we be surprised, for example, that there are no six-limbed vertebrates?

The idea of measuring disparity based on possibility rather than actuality has not escaped biologists.[3] There is a long history of biologists interested in investigating possible organic form. These include Étienne Geoffroy Saint-Hilaire, Georges Cuvier, Richard Owen, and Karl Von Baer (Eble 2000). Indeed, it has long been known that natural forms embody mathematical regularities (see, for example, Grew 1682, 152). However, despite its long prehistory, theoretical morphology is a new science; even very simple organisms demand mathematically complex models. The early development of the mathematical study of biological form is to be found in *On Growth and Form* by D'Arcy Wentworth Thompson (1917). Thompson shared both a love of natural history and a profound respect for the power of mathematics to reveal the workings of natural systems. In many respects, *On Growth and Form* is the ultimate mathematician's natural history, containing a cornucopia of mathematical facts about biological form. It also contains the first serious attempt to apply mathematics to the comparison of biological forms.

Thompson's strategy was to use Cartesian coordinates to transform one organism's shape into that of another (known as a Thompsonian transformation). The magnitude of the transformation thus represents the morphological disparity of the organisms. Figure 4.2 shows the transformation of one fish species into another by geometric deformation. Despite the originality of Thompson's approach, his methods were limited to two-dimensional analyses of species whose differences could be represented by relatively simple mathematical functions. The place of morphology in the study of evolution was reestablished with the publication of George Gaylord Simpson's *Tempo and Mode in Evolution* (1944) and *Major Features of Evolution* (1953). Simpson proposed a "new synthesis of population genetics and palaeontological approaches to the study of evolution" (in McGhee 2006, 181). But three-dimensional analysis of widely disparate morphologies required computing technology. The emergence of such technology allowed David Raup (1966) to finally model multidimensional spaces of biological possibility in ways that generated testable hypotheses.

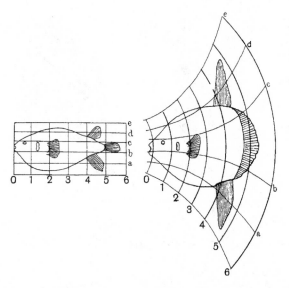

FIGURE 4.2. A Thompsonian transformation of a porcupine fish, *Diodon*, into a sunfish *Orthagoriscus*. From Thompson (1942, 1064).

4.3 BIOLOGICAL POSSIBILITY SPACES

We noted in 4.1 that Dawkins, Gould, and others have explored biological possibility with conceptual models formulated in spatial terms. Theoretical morphology's models are mathematical rather than metaphorical, and have exemplified two broad strategies. One strategy investigates the actual and possible ways a particular organism can change over time, so it employs algorithms to model changes in shape during growth and development. These take relatively simple geometrical forms and repeat them (suitably translated and/or distorted) to produce complex, seemingly biological, shapes. The output of these mathematical models can be astoundingly lifelike (as in fig. 4.3).

There is a wide variety of algorithms that can model different aspects of biological form. Many of these are best thought of as models of developmental change. These include tessellation in the patterns on a shark's skin (see, for example, Reif 1980); accretionary growth systems such as those that produce a snail's shell (see fig. 4.4); and branching growth patterns (for a survey of these and other algorithms that model botanic form, see Prusinkiewicz and Lindenmayer 1990).

Our second strategy models morphological differences between different organisms rather than the same organism at different times. These include hypothetical systems of biomechanical function such as the

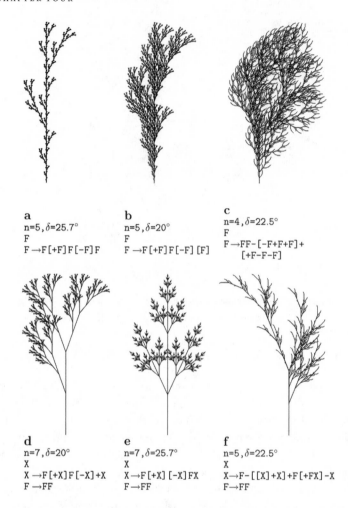

a
n=5, δ=25.7°
F
F →F[+F]F[-F]F

b
n=5, δ=20°
F
F →F[+F]F[-F][F]

c
n=4, δ=22.5°
F
F →FF-[-F+F+F]+
    [+F-F-F]

d
n=7, δ=20°
X
X →F[+X]F[-X]+X
F →FF

e
n=7, δ=25.7°
X
X →F[+X][-X]FX
F →FF

f
n=5, δ=22.5°
X
X→F-[[X]+X]+F[+FX]-X
F→FF

FIGURE 4.3. Plant forms generated using simple early Lindenmayer models of branching growth patterns. The formulas represent the generation algorithm, with δ being the branch angle and n the number of iterations of the algorithm. From Prusinkiewicz and Lindenmayer (1990). With kind permission of Springer Science and Business Media.

space of possible skeletons developed by R. D. K. Thomas and W. E. Reif (1993). Some of these are important evolutionary models that show how genuine biological novelties can gradually evolve through a series of steps, each of which (plausibly) constitutes a small improvement. Perhaps the best-known example of such a model is Dan-Eric Nilsson and Susanne Pelger's model of eye evolution (Nilsson and Pelger 1994). They construct a route from a simple light sensitive patch to a focused-lens eye. While their model is idealized, it is general enough to capture

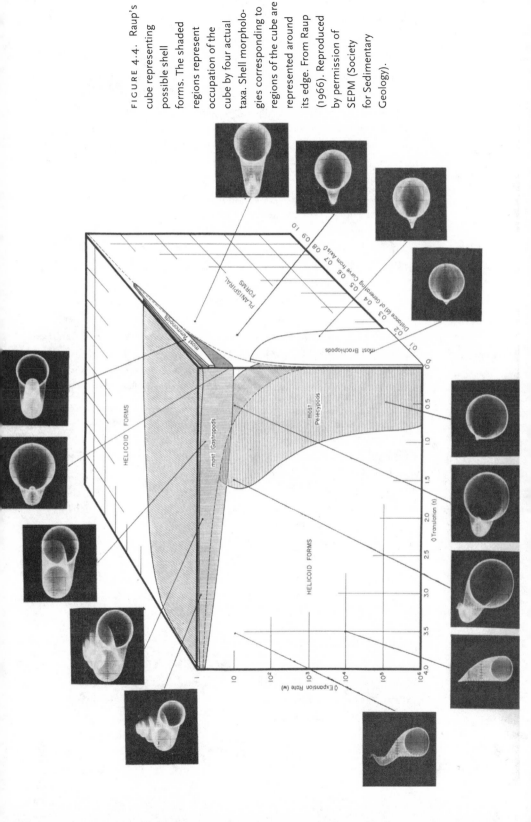

FIGURE 4.4. Raup's cube representing possible shell forms. The shaded regions represent occupation of the cube by four actual taxa. Shell morphologies corresponding to regions of the cube are represented around its edge. From Raup (1966). Reproduced by permission of SEPM (Society for Sedimentary Geology).

the structural details present in the eyes of many different organisms. Each design on the route shows some simple change from the previous one, and at each stage spatial resolution improves, by having each separate receptor cell exposed to a different field of view. The heart of the explanation is the demonstration that there is a series of gradual changes in morphology that take us from a light sensitive patch to a focused-lens eye. Models of this kind analyze complex biological mechanisms into their components, and show how small changes in these components and their interactions can enhance the mechanism of the performance as a whole. Thus they map adaptively possible trajectories through phenotype space (see Calcott [forthcoming] for a detailed account of these explanations and of their evolutionary importance).

Thus, with such models we can represent many possible biological forms, including those that are very different from actual biological forms. Even so, there are still many forms that elude the modeler's skill. There is, for example, no model that describes anything as complex as a mammal (for a good discussion of current limitations and the prospects for future advances in this methodology, see McGhee 1999, 282–87). So even with contemporary computational techniques, there are limits on the complexity of systems whose differences can be compared. Within these limits, theoretical morphology has focused strongly on spatial representations of biological possibility. The process by which such spaces are developed is simple. Parameters (length, height, rate of coiling, angle of branching, number of iterations, and so on) are deployed as dimensions of a space. These spaces thus have a dimensionality equal to the number of developmental or morphological characters studied. While, in theory, such hyperspaces might have very high dimensionality, in practice, theoretical morphologists have become adept at representing biological forms using a relatively small number of parameters. Having determined the geometry of the morphospace, it can then be used to "describe the total spectrum of physically possible forms" (Raup 1966, 1178).

It makes sense to think of a physical object being "as wide as it is deep" because width and depth can be measured in the same units. In contrast, most of the dimensions of morphospaces are only distantly related to our familiar three spatial dimensions, and their dimensions are rarely commensurable. Usually, the relationship between the dimensions is more like the relationship between the standard spatial dimensions and time. Many of the "distances" in such spaces are not even magnitudes. They are instead values within a set of discrete possible states, as in Thomas and Reif's skeleton space.

Yet in some ways these morphospaces are indeed spatial. We can "locate" actual and merely possible organisms within them. Morphospatial co-location implies similarity (with regard to the relevant parameters) just as spatial co-location implies spatial proximity. Thus by plotting the location of organisms in these morphospaces, we can reveal patterns in phenotype evolution. This process can reveal both sparsely and densely populated regions in a morphospace, both of which need explanation. We can even map the "directions" organisms take over developmental time and lineages take over evolutionary time. In the next section, we discuss a series of examples that illustrate the power of this way of representing actual and possible morphological diversity. More generally, over the next few sections we aim to show that (i) disciplined comparison between different phenotypes is indeed possible, and hence the phenotypic disparity found in a biota is both well-defined and biologically important; (ii) phenotype diversity is at best imperfectly indexed by species richness; but (iii) phenotype disparity is not conceptually independent of taxonomic diversity.

4.4 THE POWER OF MORPHOSPACES

One of the oldest and still most well known theoretical morphospaces is David Raup's cube, developed in 1966 for the analysis of shell coiling in mollusks. Raup's model rests on four parameters. Whorl expansion rate (W) is the rate at which the aperture of the shell opens out. The distance of the generating curve from the axis (D) is the rate at which the shell uncoils. The translation of the generating curve (T) is the rate at which the uncoiling shell is pushed upward from the coiling plane. The shape of the generating curve (S) is the outline of the growing edge of the shell (here assumed to be constant). Figure 4.4 represents T, W, and D in its x, y, and z axes. This is a genuine theoretical morphospace because, while it is populated by many known morphologies, it places them within a larger context of biological possibility. The space encompasses known forms, but its dimensions are not derived mathematically from any particular sample of shelled organisms. There are many potential applications for such morphospaces. The following are a few representative examples (for a more complete discussion of the value of this methodology, see Maclaurin 2003).

Morphological biodiversity is immensely important in discovering evolutionary mechanism. While evolutionary mechanisms generate both patterns in developmental mechanism and speciation, we often have access to these other consequences of evolutionary change only via their effects on morphology. Evolutionary biology is compelled to trade in inferences from morphological patterns in space and time to the mechanisms that produce those patterns. These inferences are hard to assess, even if we represent the data optimally. Hence it is important to get the patterns right: no pseudo-patterns; no pattern blindness. Hence models like Raup's are of great importance. In Raup's space, the white areas are those not populated by extant or extinct mollusks. We might therefore naturally ask (as did Raup) why these regions of possibility space have remained unpopulated. This in turn leads to an exploration of selection pressures and developmental constraints that might be responsible for the observed distributions of molluscan form. Such studies have been remarkably productive both in the discovery of morphological regularities and in the generation of novel hypotheses. We discuss one case below: the interaction of ammonite and nautilus evolution. A second common use of theoretical morphology is to plot the occupation of morphospace over time to generate and test hypotheses about the way in which clades move in response to long-term environmental change. This gives us a sense of the rates and modes of change of which different clades are capable. Mike Foote refers to this as investigating the "morphological exuberance" of clades (1997, 133). In *The Geometry of Evolution* (2006), George McGhee makes a strong case for the idea that such analysis will finally allow us to make operational the idea of the adaptive landscape.

Crucial in the success of theoretical morphology has been the contingent connection between morphological diversity and taxonomic diversity; it allows us to investigate the relationship between these two forms of biodiversity rather than assuming that species richness covaries with phenotypic variation. This is especially important in the study of major extinction events. By definition these extinguish large numbers of taxa, but employing theoretical morphology, we can investigate the relationship between taxonomic loss and morphological loss. In some cases extinction events "merely thin the number of species present, without having much effect on the total range of morphologies" (McGhee 1999, 190). In others, in cases of what David Raup calls "wanton" and "fair game" extinctions[4] (Raup 1991, 185–89), particular morphologies may be targeted in a way that will lead to loss of morphological diversity as well as taxonomic diversity.

The Cretaceous/Tertiary extinction event was both morphologically and taxonomically selective. It extinguished all the existing species of ammonites while taking a less severe toll on other similar marine mollusks, including the nautilids. This is a tale that could have been told by taxonomy. Theoretical morphology is not needed to identify the differential impact of extinction on clades, but what happened next we know about only through the efforts of theoretical morphologists. Ammonites and nautilids are both univalve marine mollusks, and so their shell morphologies can be modeled in Raup's W-D-T-S theoretical morphospace. Interestingly, the occupation of that space by nautilids takes a U-turn at the Cretaceous/Tertiary boundary as the clade wheels about in morphospace to occupy the freshly vacated territory that once belonged to the ammonites (Ward 1980). This major change in morphological trajectory is shown in figure 4.5.

But now the plot thickens; nautilids become more like ammonites, but only with respect to some of Raup's parameters. Like the ammonites before them, they are now more compressed and hydrodynamically efficient, but they never developed the highly compressed forms common among ammonites. The number of nautilid species affected makes it incontrovertible that there was selection pressure on nautilids to become more like ammonites. At the same time, fabricational constraints caused by the way the different shell types are internally buttressed prevented nautilids from wholesale annexation of ammonite morphospace. Interestingly, the ammonite territory that was not claimed by the nautilids remains vacant today, even after 65 million years of evolution.

Although the nautilids were only partially successful in exploiting the ammonite extinction, another recent discovery from theoretical

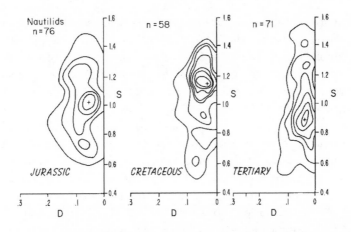

FIGURE 4.5. Frequency distribution of nautilid shell morphologies in the Jurassic, Cretaceous, and Tertiary. As in Raup's cube, D represents the distance of the generating curve and S represents its shape. Contours measure the increase in density of taxa per unit area. The morphological center of each group is represented by the +. Over the course of the Cretaceous, both the center of the clade and the majority of its members have been displaced upward, but this trend is reversed after the Cretaceous/Tertiary boundary. From Ward (1980). Reproduced by permission from *Paleobiology*.

morphology tells us that, even if ammonite species had survived, it's quite likely that the clade would never have recovered to fully reoccupy its original share of morphospace. We know this because just such a partial recovery is exactly what did happen when the ammonites rode out the earlier Permian/Triassic extinction event (McGowan 2004). It is now becoming increasingly apparent that recovery from extinction events is often not a matter of smooth reevolution. Recent work on morphology (Foote 1996; Hulsey and Wainwright 2002; Valentine et al. 2002; Roy et al. 2001, 2004; Neige 2003) causes David Jablonski to argue that the evolutionary and ecological consequences of extinction are complex as a result of "the imperfect equivalency of taxonomic, morphologic, and functional diversities" (2005, 204). Again, theoretical morphology provides us with just the tools we need to recognize the discrepancy between taxonomic and morphological recovery. Clades can recover their diversity without recapturing their disparity.

When we study mass extinctions, we ask whether phenotypic variation is lost more or less in proportion to the decline in species richness, or whether one form of biodiversity varies independently of the other. We get the same question about the relationship between diversity and disparity when species richness expands in adaptive radiations. Traditionally the famous cases such as Darwin's finches and the cichlids of Lake Victoria have been thought of in terms of speciation events giving rise to morphological disparity. Biodiversity measured by the species richness of the radiating clade, and biodiversity measured by morphological disparity were thought to go hand in hand. Mike Foote (1997) has argued that we should be careful of assuming a close fit between taxonomic and morphological radiations. Stephen Stanley argued that the Cambrian trilobites radiated and the Ordovician trilobites did not, because the recognized trilobite families expanded in the Cambrian and contracted in the Ordovician (Stanley 1990). As Foote points out, once again, morphology tells a different story. Morphometric data show that the increase in the morphological range and the variance of trilobite form was limited in the Cambrian, compared to the greater proliferation of morphological diversity in the Ordovician. That radiation occurred during a decline in family-level taxonomic diversity (Foote 1997, 134). The lesson of this example seems to be (as before) that traditional Linnaean taxonomy is a dubious compromise measure; it captures neither species richness (a decline in the number of families tells us little about species richness) nor disparity successfully.

We will close this section with a final example, Karl Niklas's model of the rapid radiation of vascular land plants. These appear as small, structurally simple forms in the late Silurian. Yet, as Niklas remarks, by the end of the Devonian, approximately 46 million years later, plants diversified phyletically and structurally to encompass all of the major plant lineages and the full spectrum of organisational grades represented in present-day floras, with the exception of flowering." (Niklas 2004, 47)

Niklas modeled this diversification by examining fitness-controlled walks through a plant morphospace. His approach combined the following three ideas.

First, he used phylogenetic information about early land plants to define the dimensions of plant morphospace and the starting position from which the lineage of land plants "explored" this space. Early plants were simple tubular structures, lacking both roots and leaves, and reproduced by releasing wind- (or water-) borne spores. The dimensions of plant morphospace require a specification of the potential branching patterns of these tubes ("axes"). This requires a determination of the density of branching (how long the tubes get before branching); the angles at which the branches diverge from one another; the angles at which one set of branches are offset from the previous branching; the symmetry of the branches; and the placement of the sites from which spores are released. The dimensionality of the morphospace is tractable and principled.

Second, Niklas assumed that selection is dominated by abiotic vectors (arguing that this was a reasonable simplification, at least in the early period of plant evolution). These were: the threat of desiccation, the threat of mechanical failure due to wind or stem overload, the need to exchange gas with the atmosphere, the need to intercept light, and the need to release spores into the wind in such a way as to ensure their dispersal. On these assumptions, the ideal plant will maximize mechanical strength, light interception, spore release, gas exchange, and water retention. If a plant had to maximize just one of these botanical virtues, well-known physical principles tell us what its ideal morphology would be. But not surprisingly, different morphologies are ideal for different virtues. Intercepting light optimally requires maximizing surface area exposed to the sun, but that accelerates water loss. Competing adaptive needs force trade-offs in plant design.

Third, trade-offs generate morphological diversity. If we take just one of these functional requirements, and search morphospace for the best designs, it turns out that only a few designs are of highest and equal fitness. If we begin a fitness controlled walk at the point of origin, our walk will lead to one of just a couple of forms. (A walk ends when we reach a point in morphospace where all the close neighbors are lower in fitness than our current location.) But as the number of demands on our plant morphologies expands, the range of equally fit forms also expands. Hence our walks can end in more and more locations. All are more-or-less equally fit, but each makes different compromises (none of these compromise morphologies are as good at any one of these tasks as a design optimal for just that task). The more demands on the morphology, the greater the distance between the all-rounder's performance and the specialist performance for each particular task (Niklas 1994; 2002).

Niklas is careful to emphasize the preliminary nature of these results. The fitness assumptions are idealized. Nutrient acquisition is left out of the model, and the assumption that each functional requirement contributes equally and independently to overall fitness is obviously unrealistic. In some environments, light energy is readily available but water is not. In others, the reverse is true. Even so, the morphologies found in these multiply constrained walks—walks in which plant morphology must compromise between all these functional demands—are similar to the morphologies found by evolution in the radiation of the land plants. This is an intriguing model, opening up the possibility of a coevolutionary interaction between ecological complexity and morphological disparity. Niklas's point about the expansion of equi-fit morphologies as distinct adaptive demands expands seems to be quite general (Marshall 2006). Nonetheless, we are agnostic about the extent to which this model explains plant diversification.

While we are agnostic about the extent to which this model explains plant diversity, we do want to underline an important element of it: the use of phylogenetic information to control our choice of dimensions for the morphospace and its point of origin. We think this constraint on the choice of dimensions is crucial to the use of morphospaces to generate testable ideas about the biological world. Likewise, it is crucial to have a principled point of origin for those models—like Nilsson and Pelger's model of eye evolution—that explore specific evolutionary trajectories. This in turn has implications for the relationship between phylogenetic diversity and morphological diversity. So far in this chapter, and in the last, we have been emphasizing the imperfect correlation between morphological disparity and species richness, even when we add in how species are sorted into clades, as in studies of mass extinction and adaptive radiation. That idea stands. But the dimensions of similarity and difference that are biologically relevant are clade-specific. In discussing phenetics in 1.2, we argued that questions of phenotype similarity and difference are ill defined unless we can answer, in a principled way: similar or different *in what ways*? Our measures of disparity are conceptually and theoretically dependent on phylogentically organized species richness. Diversity understood through phylogeny enables us to identify the similarities and differences to count. We develop this point further in the next section.

There is no doubt that theoretical morphology provides us with a new and powerful tool for the analysis of evolutionary change. The discoveries just listed would have been impossible without it. But in the final sections of this chapter we explore the power of theoretical morphology when we extend its scope beyond that of tracking closely related taxa sharing small numbers of distinct parameters. We run into its limits, for we no longer have a principled account of the dimensions of similarity and difference.

4.5 HERE THERE BE NO DRAGONS: THE LIMITS OF THEORETICAL MORPHOLOGY

In 4.4, we illustrated the use of morphospace to represent the morphological diversity of clades at and over time. These representations are mathematically tractable, theoretically principled, and biologically important. Thus in a famous series of studies, Joan Roughgarden studied the morphology of anolis lizards on Caribbean islands, comparing lizards that were alone on an island with those that shared an island with congeneric species. The idea was to test a hypothesis about character displacement. Namely, when congeneric lizards share an island they evolve away from one another, to minimize comparative interactions. Studies of this kind involve what are known as empirical morphospaces, and they form part of the wider study of actual (as opposed to possible) biological form known as morphometrics. Their dimensions are derived from observations of actual organisms. Sometimes the derivation is not direct. The dimensions of Raup's cube might not be "derived" from any particular mollusks. But clearly the power of the model stems precisely from the fact that it has been developed with molluscan architecture in mind. So these dimensions, too, have been determined by organisms. The same is true of the space of possible skeletal form developed by Thomas and Reif (1993). Here too the possibility space is based on the range of forms found, without it being closely tied to a particular clade or a particular statistical technique for detecting variation within the clade. As with Raup's shell space, this skeletal space represents plenty of merely possible morphological forms (fig. 4.6).

For reasons of both theoretical coherence and empirical tractability, it is important to have a constrained and principled choice of dimensions. More than anything else, the development of theoretical morphospaces has been a means of investigating adaptationist hypotheses. These are

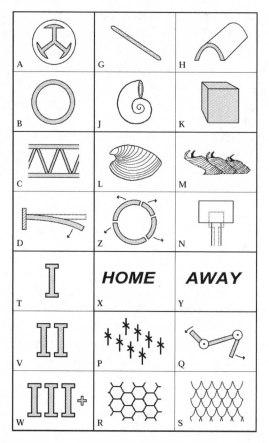

FIGURE 4.6. Dimensions of skeletal theoretical design space. The numbers be-low refer to regions of the grid representing different types of skeletal innovation. 1. Topology: either internal (A) or external (B). 2. Material: either rigid (C) or flex-ible (D). 3. Number: either one element (T), two elements (V), or three or more elements (W). 4. Geometry: either rods (G), plates (H), cones (J), or solids (K). 5. Growth pattern: either accretionary (L), unit/serial (M), replacement/molting (Z), or remodeling (N). 6. Building site: either in place (X) or in prefabrication (Y). 7. Conjunction: either in no contact (P), jointed (Q), sutured/fused, or imbricate (S). From Thomas and Reif (1993). Reprinted with permission from AAAS.

hypotheses that infer particular selection pressures from biological form and function. In this context, there is no requirement that a morpho-space represent every aspect of the morphology of the clade(s) under investigation. Ordinary morphospaces model a few parameters chosen because of their relevance to a specific hypothesis; for example, that character displacement will drive a divergence in body length. The use of these theoretical morphospaces requires that we be explicit about characteristics modeled in a particular study; that is their prime virtue. We are forced to be explicit about the pattern of change, which can then be mapped onto a phylogeny, giving us a track through morphospace, and hence empirical constraints on adaptationist hypotheses. Nobody thinks that there is some objective fact of the matter about the dimensions of lizard morphospace, independently of the hypotheses under investigation.

It is a virtue of morphospaces used for this purpose that they employ relatively few dimensions and abstract away from most characteristics of the organisms under study. But of course this also makes it clear that morphospaces developed this way are not models of overall morphological disparity. Are more ambitious uses of morphospace possible? We began chapter 3 with Gould's claim that metazoan disparity peaked in the Cambrian; we concluded that chapter with the claim that this idea was simply untestable using high taxonomic ranks to measure disparity. Could one do better by representing disparity via a morphospace? Analogous issues arise for contemporary biota. For example, a long-standing challenge to macroecology is to explain the latitudinal gradient in species richness. Tropical and near-tropical ecosystems are more densely packed with species than temperate and polar ones, even though it now appears that their speciation rates are lower (Weir and Schluter 2007). Are they also more disparate morphologically, or is tropical richness achieved (as some have suggested) by specialization, by dividing morphological and ecological space more finely?

However intriguing these questions might be, it is clear that defining a global space of morphological diversity is not an empirically or computationally tractable way of answering them. To answer Gould's question, for example, we would need a space of high dimensionality, for we would need a dimension for every respect in which Cambrian animals did vary; perhaps a dimension for every respect in which they could vary. And while we can measure some of the characters some of the time, it is clear that we cannot measure all of the characters all of the time. But even setting aside the problem of tractability, these global notions may not be well defined. How could we choose dimensions in answering Gould's question?

One option would be to use the Cambrian fauna to anchor our choice of dimensions. But that would bias our project; instead of the pull of the present biasing our taxonomic judgments about the ancient Cambrians, the limits of the ancient would bias our choice of dimensions in judging more recent disparity. Major evolutionary novelties are the invention of new ways to be similar and different. During the Cambrian radiation, the invention of segmentation, a jointed exoskeleton, and the like were not shifts within an existing space of possibilities, but the expansion of that space (Sterelny 2007). By using Cambrian animals to specify our dimension set, our representation of morphological disparity would make possibility-expanding post-Cambrian novelties invisible. And the extent and impact of such novelties is just the question at issue. If we use Cambrian, Mesozoic, and Cenozoic animals to define our dimensions, it seems likely that we will end up with incommensurable occupations of morphospace. For example, no Cambrian animals have morphological structures that make life on land possible. So if we have dimensions that reflect variation in terrestrially adapted morphologies, Cambrian animals will be unlocated (or will be assigned an arbitrary location) with respect to those dimensions. Likewise, Cenozoic animals will be unlocated on dimensions that capture the differences among the anomalocarids or *Opabinia* and its relatives. Similar issues will arise in comparing tropical and temperate morphological diversity. The choice of one anchor point will prejudge the issue; the choice of both seems likely to result in incommensurability, as tropical morphospace is likely to be more diverse with respect to some dimensions, and temperate morphospace with respect to others.

It seems to us that these global morphospaces are a morphospace too far. We think that morphospace representations of phenotype diversity are promising provided that we conceive of them as investigations of particular models based on particular biological characteristics that we antecedently believe to be important. Much work on morphospaces is of this nature. These spaces are anchored in the actual. In contrast to work using these anchored spaces, we should be much more dubious when the idea being sold is some "space of all possible." Dennett introduces one of these spaces, his Library of Mendel (1995). This is supposedly the space of all possible genomes. But he does so in part to show how problematic these conceptual zoos are.[5] Such spaces make good "intuition pumps"; they can make vivid and explicit our tacit assumptions. But we should not think these metaphors capture a well-defined and objective space of biological possibility. The boundaries of Dennett's Library of Mendel must be purely stipulative, because we have no clear conception of all possible genomes, that is, of all the ways regulation, translation, and transcription could proceed. We know what actual genes are, and we could identify some merely possible genes, in genetic systems that are close to actual ones. But we cannot go beyond this to specify all and only the possible genetic systems. This is no problem for Dennett, as his intention is merely to demonstrate the vastness of a landscape of nonactual genotypes, a task easily achieved while still leaving great swathes of possibly possible genotypes unmapped.

The Library of Mendel is bad enough; the Library of D'Arcy Thompson (the space of all possible morphologies) is worse. The Library of Mendel is at least anchored by the base sequences. We might stipulate that all possible genomes are sequences of the four bases, even though we cannot specify in advance the possible ways those sequences can be used in the generation of phenotypes, or how sequences can be grouped into genes. The morphology of organisms (or even that of animals) does not have this common currency. We cannot even specify a catalogue of all possible cell types out of which organisms could be built. Any theoretical morphospace must be anchored in actual organisms and be underpinned by some scientific goal, be it mapping the evolution of a clade, the effects of an extinction event, or the innovations in an ecological guild. It is these purposes that give morphospaces their dimensions. It is tempting to think that the availability and utility of these partial morphospaces—morphospaces that represent some aspects of the morphology of a particular group of organisms—implies that there is a global morphospace that represents all the ways organisms might have been, and locates each actual organism in that space. We think that temptation should be resisted. That space is as ill defined as the library of all possible books.

4.6 MORPHOLOGICAL BIODIVERSITY

Over the last few chapters we have been exploring the relationship between evolutionary differentiation expressed in the speciation pattern of clades and morphological differentiation. We saw in chapters 2 and 3 that these patterns are at least partly independent: it is possible for a clade to be species rich without extensive differentiation, and it is possible for a clade to be species poor, but with species markedly differentiated from their (surviving) sister taxa. For that reason, species richness is not invariably a good surrogate for morphological disparity. Even so, there is an intimate relationship between phylogenetic and morphological biodiversity. In part, that relationship is causal. We can have speciation without differentiation, and that is one reason why a clade can be species rich without being morphologically diverse. But on the Futuyma-Eldredge model (Futuyma 1987; Eldredge 1995, 2003), speciation is often necessary for morphological differentiation.

There is a second relationship as well. In 1.2, in discussing phenetics, we pointed out that similarity and difference are undefined; we must talk, instead, of similarity with respect to one or more character states. In discussing morphospace, this issue reappears as the choice of dimensions. A global morphospace is as undefined as the idea of overall similarity, and for the same reason. So phylogenetic diversity is not just part of the causal explanation of morphological diversity; it helps define the dimensions with which we evaluate the diversity of a biota. As Niklas's models of plant evolution make vivid, phylogeny gives us both our point of origin from which a partial morphospace is explored over time and a principled reason to specify a set of dimensions for such morphospaces; typically, dimensions that reflect characteristics of early members of the clade. Even here we will be selective. Our dimensions will be chosen to reflect not just any characters, but characters that matter. That choice, too, is not arbitrary, because we can build links between form and function.

The study of such links is the province of functional morphology (for a good survey of the discipline, see Plotnick and Baumiller 2000). There is considerable disagreement concerning our ability to infer function from form in fossil evidence (for a skeptical take, see Lauder 1995). Despite the methodological difficulties, even the skeptics acknowledge some successes. Lauder, for example, accepts that at the lower histological level and an upper general level of behavior and ecology we can demonstrate an important correlation between structure and function (1995, 8).[6] Of course, when we study living organisms, we are in a much better epistemic position to establish these form-function links. Physiology is obviously closely related to morphology. Disparity in shape in the wings of birds is a genuine form of biodiversity precisely because disparate shape in wing design is a good indicator of disparate aerodynamic function. Birders know the difference between the wings of falcons—narrow, pointed, tipped—and the broad, blunt-tipped wings of raptors that soar and glide. The same goes for jaw shape and bite pressure, limb morphology, hunting strategy, and the like.

In short, despite our skepticism about the idea of a global or overall morphospace, we do think that partial morphospaces capture an important dimension of biodiversity, and one that is partially independent of species richness. Moreover, these more local, anchored morphospaces allow us to make some progress in answering global questions piecemeal. For example, though Gould's hypothesis cannot be tested relative to some grand space of biological possibility, we can nibble away at it one taxon at a time. If we cannot ask whether Phanerozoic disparity has decreased, we can at least ask how often it has decreased in taxa that we have studied.

It is unlikely that we can just aggregate our results, as disparity is likely to have increased in some clades and decreased in others. Even so, the results of this investigation so far make very interesting reading. We now know that in many taxa there is a pronounced early increase in disparity. Foote's (1997, 137–38) survey of such studies finds this pattern to be widespread[7] although not ubiquitous.[8] We also know that recoveries from extinction show a wide variety of patterns. Some morphologies get replaced by subsequent radiations. Others do not. Nonetheless, there is increasing evidence showing that in some groups disparity increases after the Cambrian (Lofgren et al. 2003, 349). The jury is still out on whether there is a general trend, either up or down. Some claim a clear reduction in modern disparity (Gould 1989, 1993; Foote and Gould 1992; and Lee 1992). Others claim there is no significant difference in disparity between recent and Cambrian arthropods (Briggs et al. 1992a; 1992b).

Even if there is not much change in total arthropod disparity, there has been a change in the regions of morphospace occupied. Crustacean and trilobite-like forms dominate Cambrian disparity. But by the Carboniferous, chelicerate forms dominate, and modern disparity is correspondingly dominated by the insect-dominated hexapods (Lofgren et al. 2003). This migration through morphospace is seen in many clades and at many scales, as is evident from Lofgren's own study of hexapod disparity (see fig. 4.7).

It is time to tie together the discussion of the last two chapters about the relationship between species richness and morphological difference

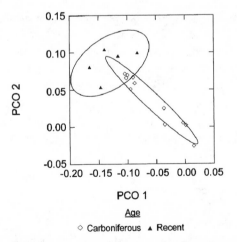

FIGURE 4.7. Distributions of hexapods from both the Carboniferous and the Recent. PCO 1 and 2 are the principal coordinates, statistically derived, of variation within the sample. While there is considerable disparity in both ancient and modern hexapods, there is relatively little intersection between the regions of morphospace they occupy. So the morphology of most ancient species was very different from that of their modern descendants. From Lofgren et al. (2003, 360). Reproduced by permission from *Paleobiology*.

across a clade or group of clades. Species richness and morphological difference are not the same thing, but nevertheless, it might well be the case that species level diversity tracks and in part explains morphological diversity. We saw in chapter 3 that Gould and his paleobiological allies do not think so, but we have also seen that paleobiology had no good framework in which to formulate and defend their intuitive judgments about the morphological disparity of Cambrian animals. Moreover, there is a natural explanation of those judgments in terms of the distinction between stem group and crown group taxa.

Chapter 4 was focused on this framework problem. There is a program for representing morphology and morphological difference as a morphospace. If a morphospace is, as it seems, a way of representing all possible varieties of organic form, this way of representing morphology would enable us to decouple claims about the morphological disparity in a biota from claims about species richness and phylogenetic organization. We can represent patterns of phenotype evolution independently of issues of phylogeny and species richness, and then assess the extent to which the first varies independently of the second. We have argued that this strategy is successful, but only in limited ways. We must choose the dimensions of a morphospace in principled ways, and we must also (often) make a principled decision about the point (or points) in that space from which lineages begin their evolutionary explorations of evolutionary possibility. These requirements make it necessary for us to anchor morphospaces in actual taxa and their histories. We can still use them (as we have shown) to explore the ways in which species richness and morphological diversity interact in a clade: we can represent the expansion of structure in the vascular plants; the survival of lineages through mass extinction events, but with a permanent loss of morphological diversity; and the inaccessibility of much of shell space. The construction of a morphospace is a powerful representational tool for exploring phenotype evolution and its relation to speciation patterns. Furthermore, this use certainly reinforces the intuitive expectation that patterns of phenotype evolution do not perfectly track speciation patterns.

However, its use is local and limited. Unless we have a specific clade and explanatory project in mind, there is no principled answer to the question "What are the dimensions of that morphospace?" It follows, then, that the question of whether species richness (measured just in sheer numbers, or with phylogenetic information) covaries well with morphological and phenotypic diversity has no fully general answer. For issues of evolutionary biology, we clearly cannot use phylogenetically structured species richness information as a surrogate for morphological information; many of the explanatory projects of evolutionary biology concern the interaction of speciation and phenotype evolution. Hence phenotype change has to be tracked separately. But the aspects of phenotypes that are tracked are, as we have noted, often anchored by phylogenetic information (for example, information about primitive and derived features). Likewise, ecologists and conservation biologists will sometimes be interested in the phenotype variation found in a biological system. In considering whether a grassland is resilient in the face of disturbance, it might not matter whether we have a few species with a lot of phenotypic variation, or many species each with less variation. For determining the resilience of a grassland confronted with (say) a drought, the crucial issue will be the availability of specific phenotypes that minimize water loss, maximize water use, or that enable the plant to persist in a dormant state. In contrast to these cases where a specific, known threat enables us to identify specific aspects of phenotype, conservation biologists and ecologists will sometimes be interested in the resilience of a biota in the face of changes whose kind, shape, and intensity are not known in advance. So they will often be able to treat phylogenetically structured species information as a surrogate for the spread of phenotypes in a biota.

5   *Development and Diversity*

5.1 DIVERSITY, DISPARITY, PLASTICITY

Biology is a historical science, and one of its most important projects is to predict and explain change. To put it mildly, this project is not merely of theoretical importance. Climate change is upon us, and if possible we need to anticipate the responses of local, regional, and global biotas to the array of environmental perturbations likely to accompany climatic change. In this chapter we take up the critical question of whether populations respond to these changes individually, according to their own habitat preferences, or whether their response is modulated by their interactions with their neighbors, mediated by the structure and organization of ecological systems. Clearly, too, the standing phenotypic variation—the diversity of available phenotypes—plays an important role in explaining biotic response to environmental change. If a regional biota already contains organisms whose phenotypes make them storm resistant, or drought resistant, or able to stabilize soils in the face of increased runoff, we can expect a buffered response to change. We might see some changes in abundance and membership, but not a wholesale reconstruction of regional systems and biotas. That is one reason why disparity matters; all else equal, the more disparate the biota, the greater its standing morphological-phenotypic variation, the more resilient it will be in the face of change.

We argued in chapter 2 that phylogenetic structure matters too. It is important to identify how standing phenotypic-morphological variation is parceled out into species, because locally adaptive variations that are not protected by reproductive isolation may not be robust in the face of ecological change. If ecological changes remix the populations, some of the standing phenotypic variation in a widespread species with local varieties could well disappear. That will not happen if the local variations are entrenched by speciation. In chapters 3 and 4, we projected this phylogenetic structure into deep time, to ask whether, in general, evolutionarily deep, species-rich clades are morphologically varied as well, or whether we should think of morphological diversity as varying independently of species richness. This issue turned out to be far from straightforward. Dawkins's "genetic space" is seductive but unhelpful. While morphology does vary independently of species richness, the dimensions of variation have to be identified in the light of the evolving history of the lineage. While local morphospaces of low dimensionality turn out to be an important conception of biodiversity, we argued in 4.6 that the dimensions of that space depend on the clade (or clades) whose fate we want to explain or predict. But how, exactly? We answer that question in 5.4. The developmental system of a lineage determines those aspects of phenotype that can vary independently, and it is these that we represent through our choice of dimensions.

In this chapter, we will be considering a fourth factor: hidden biodiversity. Variation between both populations and individuals can be morphologically invisible. Two populations can be phenotypically similar yet vary in their genetic resources or in the distribution of those resources. Two individuals can be phenotypically similar, yet differ in their developmental biology. The lineages that those individuals represent might therefore have very different fates or potentials. So in part, this chapter continues the argument of chapter 4. It is a further development of the project of supplementing a phylogenetically informed species richness measure of biodiversity with a tractable and principled concept of morphological diversity. In part, it develops the case for a second supplement, one specifying the evolutionary potential, and the differences in potential, of clades. In 4.5 and 4.6, we argued that it is important to formulate a coherent and tractable concept of morphological diversity, even if morphological diversity is well indexed by species richness, even if conserving one conserves the other. We need a good measure of morphological disparity to investigate the ecological and evolutionary interplay between diversity and disparity. Indeed, we can only know that diversity tracks disparity if we can independently measure each. In 5.5, we shall develop a similar suggestion about evolutionary plasticity: to tell whether plasticity covaries with diversity and disparity, we need to be able to represent it independently of diversity and disparity.

We begin with an example that makes the difference between morphology and population structure vivid, and that illustrates the importance of that population structure. Reproductively isolated populations of Australasian robins (genus *Petroica*) are phenotypically similar, but differ greatly in their ability to respond to environmental change. The black robin of the Chatham Islands (*Petroica traversi*) now has a population size of about 250, but in 1980 it was reduced to a single breeding pair. Color aside, the black robin is phenotypically very similar to its New Zealand and Australian congeners. But, like many threatened species that have passed through population bottlenecks, it has been stripped of much of its genetic variability. This has undoubtedly cost it dearly in evolutionary plasticity. It now has a diminished capacity to respond to environmental change. It is just this phenomenon that underpins the "small population paradigm" that has dominated much of conservation biology (Caughley 1995, 216–27).[1]

So conservation biologists have good phylogenetic reasons for sampling gene pools: consistent genetic differences of suitable magnitude among populations are often good evidence of cryptic speciation; reproductive isolation has allowed gene pools to diverge.[2] But there are also good reasons to monitor variation within populations, too, reasons to do with the evolutionary resilience of vulnerable species in the face of change and the genetic load imposed on these populations by forced inbreeding (we return to these issues in chapter 7).

The upshot of this line of thought is that there is a causally consequential but hidden dimension of biodiversity: the genetic variability of a population or species. Conservation biologists have typically been interested in population-level properties of a species and their consequences: the size of population, its fragmentation and metapopulation dynamics, and age and sex structure. Such population-level properties are important. For example, island populations of birds are vulnerable to extinction in part because of these population-level properties. Populations are typically small (which both reduces variation and makes them more likely to fluctuate to zero as a result of a bad season) and demographically isolated (hence much less likely to be rescued by migration). Local extinction *is* extinction. But this dimension is of interest not just to conservation biologists modeling short-term change in small populations. It is important as well in the literature on developmental and phylogenetic constraints, as it influences the routes available through morphospace from a lineage's current location. However, for those interested in developmental and phylogenetic constraints, the interest shifts to include not just population-level properties but also the properties of individual organisms. To stay with birds and islands, one striking pattern in bird evolution is the repeated evolution on islands of flightless rails (many now extinct) (see Trewick 1997; Steadman and Martin 2003). Other birds have become flightless: a few ducks and their relatives, a few passerines, and a Galapagos shag. Even so, the loss of flight is common in the rail clade and comparatively rare in other lineages, and that pattern requires explanation. This example suggests that clades of roughly equal evolutionary depth and species richness can vary in their evolutionary plasticity.

As we see it, then, developmental differences between lineages are important because they contribute to differences in evolutionary plasticity (or "evolvability") on both short and long time frames. The differences can derive from differences in *standing variation* but also from differences in *accessible variation*: variants that are likely to arise given the population structure, environment, and developmental biology of the species. Genetic variability[3] is of particular significance to conservation biology, but only because it's an important contributor to plasticity that can easily be lost as populations shrink. In the next section, we sketch the range of developmental resources (and hence differences in those resources) that contribute to evolutionary plasticity. We then illustrate these issues by discussing a salient case in more detail, namely developmental modularity, currently the hottest of hot topics in evolutionary developmental biology. Once again, granting that the units or elements over which biodiversity is defined are species, what are the salient differences and similarities? The idea is that differences in both population organization and developmental biology (differences that may not be echoed in morphology) are relevant to biodiversity measures of a biota, in both short-term conservation contexts and long-term evolutionary contexts.

5.2 THE VARIETY OF DEVELOPMENTAL RESOURCES

It is a notorious, and much discussed, feature of the modern synthesis that it neglected developmental biology. There was no *Embryology and the Origin of Species* to join *Genetics and the Origin of Species*, *Systematics and the Origin of Species*, *Tempo and Mode in Evolution*, and *Variation and Evolution in Plants* as core elements of the evolutionary consensus of the 1950s and 1960s. Through this period population genetics was the central synthesis discipline; indeed, in the eyes of many, population genetics was evolutionary biology. Hence the infamous definition of evolution as change in gene frequency. The neglect of developmental biology made some sense in the light of the working assumptions of the time. It was supposed that that variation in natural populations was typically distributed densely and without bias around its current mean. If that assumption were satisfied, we could predict the future of a lineage, knowing only the volume of morphospace it now occupied and its current and future environments, for these would determine the impact of selection on its trajectory. We would not need additional information about available genetic resources and the mechanisms that use them, for developmental mechanisms affect evolutionary trajectories only if they result in a biased or restricted flow of variation to selection. If they do not, evolutionary biologists can reasonably idealize away from the complexities of development and treat the current morphology of a lineage as a good guide to its future possibilities.

However, we no longer have good reason to suppose that this idealization is appropriate. It was tied to Ronald Fisher's models that supposed that the development of a phenotypic trait (and hence variation in that trait) depended on a large number of small-effect genes. But mutation is not restricted to point mutations substituting one amino acid for another. Mutation can result in movement, duplication, inversion, and deletion of DNA sequences, and hence can result in changes to gene regulation and to shifts in reading frames, as well as changes to the amino acids that are read off DNA sequences. Some of these less usual genetic events are likely to be important in many evolutionary changes. Their phenotypic effects need not be distributed normally around current phenotype values (for a succinct and elegant review of these issues, see Orr 2005).

At the end of 5.1, we suggested that accessible variation is important to explaining the evolutionary possibilities open to a lineage. In our view, accessible variation depends on three sets of factors: (i) One set of factors are those that influence the nature of novel gene combinations and the rate at which they are formed. (ii) Genetic novelties once made can be lost. So a second set of factors that influence the rate at which novel genes and gene combinations are accumulated in a lineage are important. (iii) New gene combinations vary in their phenotypic influence. So the third set of factors influence the nature and rate at which novel phenotypes are formed from these new resources. In this section, we sketch these ideas; in the next, we explore one example in some detail.

Adding Genetic Novelties

Lineages differ from one another in recombination and mutation rates. Indeed, the most obvious factor responsible for adding genetic novelty is, of course, mutation. For a given mutation rate, large populations will find nearby novelties more rapidly than small ones, so population size matters. So too does the mutation rate itself, which varies. As the molecular clock debates have made clear, rates differ across lineages and for different kinds of mutation. Even within a lineage, the mutation rate is not constant across the genome. But novelty often consists in novel gene combinations, and that makes population structure important as well. Population structure distributes the genetic resources of the species to its subpopulations. Microevolutionary change takes place within local populations, and if these are isolated from one another, there may well be potentially important gene combinations that are unavailable, because the elements that would form the combination have arisen in different populations. So the extent to which a population is divided into somewhat varying subpopulations and the migration rates between those subpopulations are relevant to the rate at which novel gene combinations form. For the same reason, the mating system is also important. Some mating systems impede the flow of genes across the population, and others promote it. Population structure is even more obviously important for prokaryote populations (O'Malley and Dupré 2007). Prokaryotes have limited chromosomal evolution (their chromosome is circular, so there is no recombination). But there is rich horizontal transfer of ready-made genetic material. Plasmids, phage DNA, and transposons are all mechanisms of horizontal gene movement, of different size packets. Given the ubiquity of horizontal gene transfer, the richness of the pool of local genetic resources is obviously important (Carroll 2002).

Accumulating Genetic Variation

Once novel genes and gene combinations are found, they might be retained, amplified, or lost. The environment plays a role in promoting or impeding accumulation. Heterogeneous environments sometimes preserve variation, because different gene combinations are favored in different environments. Somewhat counterintuitively, a homogenous environment can also keep genes in the population by keeping them phenotypically equivalent. In many cases, the expression of a gene is context sensitive. Fix an allele, fix its genetic context, but vary the environment and often the resulting phenotype varies. This context-sensitivity of gene action is often represented by a curve that plots environmental variation against phenotype variation, while holding genetic factors constant (see fig. 5.1). The phenomenon itself is known as a norm of reaction. Importantly, the norm of reaction of different alleles can coincide in some but not all environments. In particular, genes quite often have equivalent effects in typical environments but different effects when the organism is under stress, or when it develops in unusual circumstances (see Schlichting and Pigliucci 1998). Uniform environments can allow variation to accumulate by masking its expression, because the environment a lineage experiences determines the fraction of the reaction norm that is expressed and thus exposed to selection. More uniform environments

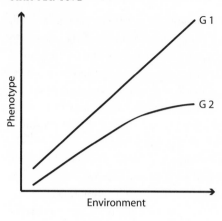

FIGURE 5.1. Reaction norms for two genotypes with respect to two hypothetical environmental and phenotypic parameters.

allow cryptic genetic variation to survive unexpressed, variation that may become important if the environment changes.

Suzannah Rutherford (2000) develops this point in some detail. Populations have unexpressed genetic variability; for example, in natural *Drosophila melanogaster* populations there are hundreds of thousands of base pair differences between the haploid genotypes. Yet these are strikingly uniform populations. Much genetic variation is effectively neutral because it does not give rise to phenotypic variation. Although this variation is cryptic while the environment is stable, it can be unmasked. It is *unexpressed* difference, not *inexpressible* difference. Consider, for example, the two forms of the human *CYP1A1* gene. In non-smokers, these two forms are phenotypically equivalent. But they are associated with marked difference in lung cancer risk for smokers. In particular, one form of the gene makes moderate smoking much more dangerous than it would otherwise be (Rutherford 2000). Uniform environments thus allow genetic variation to be stored. They enhance long-run evolvability by preserving genetic variation that would otherwise be eliminated from the gene pool. This uniformity effect may well be quite important. Many organisms act on their environment in ways that homogenize them. For example, animals that mature in nests, burrows, termite mounds, and the like experience a relatively uniform environment. Temperature, humidity, and other environmental fluxes are controlled (Odling-Smee et al. 2003).

The Use of Genetic Resources and the Genotype-Phenotype Map

The supply and accumulation of genetic variation is obviously relevant to the evolutionary plasticity of a lineage. But it is not the only factor that matters. Of importance is the use of genetic resources by the developmental machinery inherited by a lineage.[4] This is an enormous topic, for it is the core of developmental biology, and so we can only touch on it here. In the rest of this section, we will discuss important new ideas about the relationship between developmental and evolutionary plasticity. In sections 5.3 and 5.4, we discuss developmental and evolutionary modularity: the idea that certain characteristics of organisms develop independently of other traits, and hence can evolve independently of those other traits. On this view, organisms are developmental mosaics. This is the fundamental idea of evolutionary developmental biology. It links us back to the themes of 1.2, 4.5, and 4.6. There we struggled with the problem of identifying the relevant similarities and differences between organisms. The mosaic hypothesis of evolutionary developmental biology offers a principled solution to that problem: the structure of the developmental mosaic characteristic of a clade corresponds to the real dimensions of morphospace for that clade. This places limits on the size of clade that can be successfully subject to morphological comparison, hence it gives us a practical maximum size for a partial morphospace.

We begin with the idea that there is a profound connection between developmental and evolutionary plasticity (for early and important work on this, see Schlichting and Pigliucci 1998). The existence of developmental plasticity itself is not controversial; one way it can be represented is through a reaction norm of the kind discussed above. Nor is it controversial that developmental plasticity is often adaptive. For example, in growth, skeletons respond to physical stress by increasing the strength of load-bearing elements. What is novel and controversial is the idea that developmental plasticity is central to evolutionary plasticity. This idea and its evolutionary consequences have been explored by Mary Jane West-Eberhard and, in collaboration, John Gerhart and Marc Kirschner. They have all recently argued that within-generation plasticity is a preadaptation to evolutionary plasticity. Lineages are evolutionarily plastic because organisms are developmentally plastic (Gerhart and Kirschner 1997; West-Eberhard 2003; Kirschner et al. 2005).

Kirschner and Gerhart begin with the observation that adaptive phenotypic plasticity is essential for complex organisms. The organization of a complex organism cannot be controlled precisely by inherited information; there cannot be a complete genetic specification of a phenotype. Developing embryos will be exposed to differing environmental fluxes, and will be supplied with differing nutrient packages. These will have affects on developmental trajectories, and so a given component needs to be able to work in somewhat different internal environments. Its systems of signaling, coordination, and linkage must be able to cope with somewhat varying organizations of internal components. The exact shape, location, and structure of these components cannot be predicted in advance, yet organ systems must be appropriately connected to one another if the organism is to function. For example, Kirschner and colleagues (2005) point out that the vascular system of mammals is extraordinarily complex; cells are never more than a few cell diameters away from a capillary. Yet the precise plumbing cannot be prespecified; it must be sensitive to bone and muscle growth. Capillaries are provided by a process of oversupply and selective attrition. Wherever muscle development is dense, and hence the need for oxygen flow is great, less of the oversupply will be deleted. This seems to be a common pattern among the mechanisms that adjust one system in response to contingencies elsewhere.

In West-Eberhard's treatise on developmental plasticity, she discusses the surprising power of many such mechanisms. For example, in human populations there are many pathologically developed hearts with arteries, veins, and valves in nonstandard places. While these may not be optimal, these developmental pathologies are not instantly fatal. The rest of the phenotype accommodates to them, connecting the system functionally to the circulatory and respiratory system. These mechanisms are of great evolutionary significance, where they exist, because they make possible phenotypic adjustment to genetically driven novelty elsewhere in the phenotype. Adjustments will not require correlated genetic change. If sexual selection increases the neck muscle mass of a male deer, there is no need for further genetic changes to ensure that those muscles are adequately serviced by the circulatory and nervous system of the animal. Similarly, the mechanism in mitosis that ensures that each daughter cell receives the right chromosome complement (spindle formation) is adaptively plastic. It is not and cannot be preprogrammed with the location of the chromosomes in the dividing mother cell. So the microtubules that usher them to the daughter cells explore from the centriole, and are stabilized if they connect with a chromosome. If they do not, they are reabsorbed, and new microtubules form (Kirschner and Gerhart 1998). Hence, mutations that increase chromosome number need not be fatal. Without a mechanism that adjusts one structure in response to changes in another, a coordination problem would severely constrain adaptive change. Phenotypic accommodation reduces the problem of correlated change. Genetically caused modification in one system need not wait for a genetically caused change in associated systems, even when both organ systems must change for either change to be adaptive.

So mechanisms of phenotypic plasticity enhance evolvability by enabling phenotypic adjustment to genetically caused changes in an organism. These mechanisms act as *change amplifiers*. Genetic changes that directly affect only one component of an organism can result in a suite of adaptively correlated changes. Thus a small genetic change can map onto a large phenotypic change, via these knock-on effects.

Plasticity enhances the flow of variation to selection. Thus West-Eberhard argues that macroevolutionary differences between major clades are explained by differences between those clades in their inherited mechanisms of developmental plasticity.

In discussing the classic evolutionary problem of adaptive radiations, she calls this the "flexible stem" hypothesis. As she notes, received wisdom on adaptive radiation sees it as driven by ecology. Populations from a stem species radiate into a diverse set of niches. Selection acting on them differentially leads to phenotypic divergence, typically accompanied by speciation (see Schluter [2000] for an extensive discussion of this model). West-Eberhard points out that this cannot be the whole story; in the classic examples of radiations on islands, some of the founding migrants are stem species of adaptive radiations, but many are not. She argues that adaptive radiations take place when a flexible stem species provides migrants for a diverse set of environments (see West-Eberhard 2003, chap. 29). As she sees it, the flexible stem species must be developmentally plastic, thus allowing the migrants to survive by phenotypic adjustment to their new circumstances. But the flexible stem must also be flexible in evolutionary time, as those phenotypic adjustments are amplified and entrenched genetically. She suggests that the über-example of adaptive radiation, that of the cichlids in Lake Victoria, exemplify this pattern. The cichlid radiation was largely in feeding morphology, a radiation made possible in part by modularity and redundancy in the cichlids' double set of jaws. But cichlid feeding morphology is developmentally plastic, too. Given different diets, cichlids of the very same species develop different tooth and jaw morphologies. (All this might be further enhanced, she speculates, by sexual selection on jaw anatomy, so that differences in feeding morphology between populations leads them to become reproductively isolated as well.)

5.3 FROM GENE REGULATION TO MODULARITY

We think these new ideas about the evolutionary role of developmental plasticity are important. But they are recent, and, as we noted above, the main focus of discussion has been on the so-called genotype $\Rightarrow$ phenotype map, and in particular on modularity. There are in the literature a number of concepts of modularity, and they are not equivalent (see Box 5.1). But, roughly speaking, a trait is modular if its development is relatively independent of the development of the other traits of the organism, so perturbation in the modular trait does not result in perturbation of others.

BOX 5.1: Different Concepts of Modularity

This taxonomy of modularity concepts roughly follows the analysis of Wagner and Mezey (2004).

Modules as morphology

Regulatory genes can be manipulated to build normal structures in abnormal places. So one concept of modularity is tied to the idea that we demonstrate modularity by experiments that induce ectopic development. In this sense, a module is "any developmentally autonomous part of the embryo that can develop all or most of its structure outside its normal context" (Wagner and Mezey 2004, 339). A fly's leg is thus a module by this definition because we can experimentally induce the production of legs at nonstandard locations. This notion of a module is thus both developmental and structural; a module is an autonomously developing morphological unit.

Modules as developmental cascades

Viewed in terms of developmental processes, a module consists of a set of developmentally downstream elements marked out by upstream choice points; see, for example, West-Eberhard (2003, 54). These modules are relatively autonomous, identifiable cascades. Since the same genes are available in all the cells of an organism's body (and as homologies in related clades) the one developmental switch can be used to initiate cascades leading to quite different structures; for example, to morphologically distinct limbs. Ancient genes such as *Eyeless* and *Tinman* mark genetic process modules as does the mammalian gene *Hoxa-11*. This is used variously in the development of limbs, kidneys, and the female urogenital organs. Genetic process modules may be developmental cascades downstream from single genes or gene networks and, unlike developmental modules, they may perform quite abstract tasks such as making or maintaining an asymmetric boundary (Wagner and Mezey 2004, 340).

Modules as variants

As noted at the start of this chapter, the plasticity of a lineage rests on its ability to enhance and utilize its standing phenotypic variation. So a third way to think about modules is as elements of independently selectable variation. It is modularity in this sense that fulfils Lewontin's demand that natural selection must act upon traits that are quasi-independent of one another. Modules in this sense need to be more than identifiable developmental cascades. They must also be endowed with a separable function, one that can make an independent contribution to fitness (Schlosser and Wagner 2004).

In different ways, and using different language, Lewontin, Dawkins, Raff, Wimsatt, Kauffman, and Wagner have all explored the idea that the evolutionary plasticity of a trait depends on the extent to which it can vary independently of other traits.[5] For example, Stuart Kauffman (1993; 1995) argues that adaptive evolution is possible only if small changes in genotype typically cause small changes in phenotype, and these in turn typically cause small changes in fitness. The key claim is that adaptive response to selection is possible only if, and to the extent that, fitness-relevant characteristics are, in Richard Lewontin's terminology, "quasi-independent": they can change independently of other aspects of the organism's phenotype. For example, in his *Biased Embryos and Evolution*, Wallace Arthur conjectured that among mammals, limb length is not quasi-independent; the mechanisms that generate selectable variation rarely generate left/right or front/back length asymmetries (Arthur 2004). It is clear from this example (and from the existence of kangaroos, kangaroo rats, and fiddler crabs) that quasi-independence is a degree concept. Traits are *more or less* independent of one another in their developmental trajectories, and hence, as we would expect, the evolutionary plasticity of a trait is a matter of degree. Interestingly, the connection between evolvability and modularity is experimentally supported by a nonbiological example of modularity in development (see Box 5.2).

BOX 5.2: Evolutionary Computation

The importance of modularity has been reinforced by evolutionary computation theory; the study of evolutionary algorithms. These are computer programs that are subjected to mutation (small random changes in code) and recombination with other programs in a candidate population of programs. Their "fitness" is tested by comparing their performance in a prespecified task, for example, producing a jazz solo judged acceptable by experimental subjects (Biles 1994). Early experiments in the field (see, for example, Friedberg 1959) involved randomly altering the source code of ordinary computer programs and attempting to select the successful offspring. These early experiments were unsuccessful—to put it mildly.

Quite small changes in the source code produced "offspring" programs that either would not run at all or performed in a manner that was massively unpredictable. Perhaps that is not so surprising to those of us familiar with the potentially disastrous results of "tinkering" with the code that runs a personal computer. However, computer programs can be written in such a way that they do exhibit greatly enhanced evolvability. While other factors are important,[6] what all successful "genetic algorithms" have in common is that they are highly modular both in the sense that they produce standard outputs from standard inputs and in the sense that there is tight integration within modules but relatively little integration between elements of different modules. This produces a usable proportion of viable "offspring" that have been used to produce solutions to complex computing problems such as the deployment of face recognition software (see, for example, Caldwell and Johnston's 1991 "Tracking a Criminal Suspect through 'Face-Space' with a Genetic Algorithm"). Such algorithms are similar to biological instances of modular development in other respects. For example, genetic algorithms pass Kauffman's test: small changes in the underlying code typically cause small changes in phenotype, and these in turn typically cause small changes in the program's performance.

Modularity contrasts with generative entrenchment, a concept introduced by William Wimsatt to describe mechanisms and traits that are connected in development to many features of an organism's phenotype. In his view, mechanisms that evolve early and develop early in ontogeny become increasingly resistant to evolutionary change (as in fig. 5.2) because they are increasingly implicated in many ontogenetic processes (Wimsatt and Schank 1988; Wimsatt 2001, 2007). The gene code—the mechanisms that translate and transcribe the DNA itself—are perhaps the clearest example. These are causally implicated in all gene expression. So it is hard to imagine selection in favor of a mutation that, say, made development less sensitive to the stop codon. Such a mutation would have effects on so many episodes of gene activity that something would be bound to go horribly wrong.

In virtue of their broad developmental relevance, a change in early systems is likely to have some appalling consequence somewhere in ontogeny.[7] This line of argument makes it plausible to think that the body organization that characterizes a phylum is entrenched and hence evolutionarily inflexible. This morphological organization evolved early[8] and typically, though not invariably, appears early in ontogeny (the phylotypic stage—see Raff 1996). The basic body plan of, say, an arthropod—the idea goes—is laid down early in development and scaffolds the further development of the specific organizational and organ systems that char-

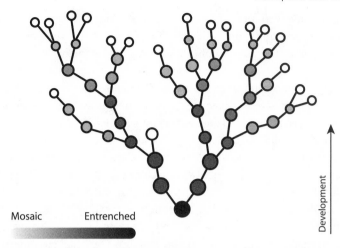

FIGURE 5.2. Generative entrenchment. The entrenchment of each module in this hypothetical developmental cascade is a function of the number of modules that are developmentally downstream from it.

acterize particular kinds of arthropod: the beetles, the shrimps, the spiders, and so on. On this model developmental entrenchment explains why the Cambrian radiation was uniquely disparate and why Mesozoic and Cainozoic evolution has involved variation upon a relatively small number of themes (see also McShea 1993).

Lewontin, Wimsatt, and Kauffmann were led to the importance of modularity through evolutionary theory. In "Adaptation," a justly celebrated paper (1985), Lewontin (for example) argued that populations could respond to selection only on traits that were "quasi-independent." Such traits can change without the rest changing, and these can respond to selection. These theoretical considerations proved congruent with striking experimental results in developmental biology. It turned out to be possible to induce complete organ systems to develop in the wrong place. This result suggested that the organ-building cascade in question was self-contained. Evolutionary considerations were sustained by the discovery of systems of gene regulation, of genetic switches that cause relatively autonomous developmental cascades.

These were first discovered in prokaryotes, and analogous mechanisms have subsequently been discovered in more complex organisms. François Jacob and Jacques Monod's (1961) discovery of genetic switching in *E. coli* gave us the ability to explain cell differentiation in a wide variety of organisms (Carroll 2005, 60). Subsequently this has been used to explain single gene mutations in fruit flies that alter the number and placement of body parts. These "homeotic" mutations produce complete and well-formed structures in the wrong places. Thus the so-called bithorax mutations produce a four-winged fly in which the third segment of the thorax is a replicate of the second segment, the segment carrying the two wings of the wild-type fly.

The still more bizarre Antennapedia mutations produce flies in which legs replace antennae (fig. 5.3). These are striking examples of the actions of regulatory genes. Aberrant genetic signals initiate a cascade that is completed despite the abnormal location of these processes in the insect's body. The cascade proceeds relatively normally despite its unusual location, from which we can infer that it is essentially under local control. The structures depend on gene regulation systems that control developmental modules, autonomous developmental cascades. Moreover, if functional eyes and antennae can be produced outside their normal developmental context, then it is reasonable to infer that the normal development of legs, wings, and eyes is also developmentally disconnected from the surrounding tissues. In this way, the new developmental biology has given us a whole new explanatory framework for decomposing organisms into discrete traits.

These homeotic mutations are central to our understanding of evolution, because the genes responsible for them, the Hox genes, are much more widespread than was originally anticipated (McGinnis et al. 1984). Furthermore, the Hox genes of insects, worms, frogs, and mammals are largely homologous (see fig. 5.4). Hox genes are shared in some version by all Metazoa. They play an important role in front-to-back dif-

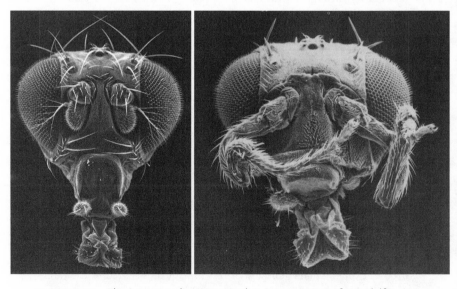

FIGURE 5.3. The Antennapedia mutant (right). By permission of F. Rudolf Turner.

Drosophila embryo

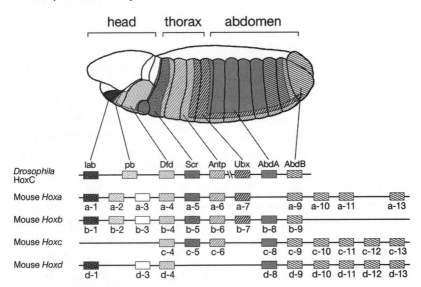

Mouse embryo

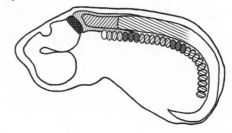

FIGURE 5.4. Hox gene expression and organization in the embryos of a fruit fly (above) and a mouse (below). Shading shows the regions of the developing embryos in which each gene is expressed. By permission of Sean Carroll. Drawn by Leanne Olds.

ferentiation and segment identity. So we know that there exist ancient, conserved, and widely shared regulatory genes that seem to initialize more-or-less autonomous developmental cascades, and subsequent investigation has extended this phenomena beyond the Hox genes. *Eyeless* in flies, now known as *Pax-6*, has turned out to be homologous with *Small Eye* in mice and *Anaridia* in humans. Homologues of *dll* (originally *distal-less* in flies, as its mutated form causes the loss of the distal parts of the fly's limbs) have been found in organisms as disparate as chickens, fish, and sea urchins. *Tinman*, which triggers the formation of the fly's heart, has turned out to be a homologue of the mammalian *NK2* gene family.

On a broader scale, recent advances in gene sequencing show that moderately closely related organisms share a huge proportion of their genomes. The fact that morphological disparity seems to co-occur with overall sequence similarity leads Sean Carroll to infer that evolutionary history is largely the history of changes, not in genetic makeup, but in gene regulation (2005, 270). In short, recent advances in developmental biology make it likely that genetic switches stand behind the production of organ systems, organs, limbs, and coloration patterns—among many other functional characteristics. They regulate the placement, timing, and quantity of these biological traits. They often act independently of other switches. Thus there is suggestive evidence that important aspects of the developmental biology of a clade is conserved, sometimes for very long periods of time, even though the clade has diverged morphologically. The conservation of Hox mechanisms in the bilaterian animals is the flagship example of this phenomenon. If the phenomenon is widespread, we can legitimately speak of the plasticity of lineages. And there is suggestive evidence that the development of important structural features of organisms is under the control of autonomous genetic switches. Eye development is the flagship example of this phenomenon. If it is general, we can legitimately link developmental modularity with morphological organization.

5.4 MODULARITY IN DEVELOPMENT AND EVOLUTION

We noted above that there are a number of concepts of modularity floating around in the literature, some more focused on development, others more explicitly evolutionary. Evolutionary and developmental considerations do not pick out quite the same traits. The evolutionary notion is of a trait that reacts to natural selection as a unit, of traits that are quasi-independent of one another. But while developmental autonomy of the kind that is captured by experimentally induced development in abnormal places might be necessary for naturally occurring and selectable morphological variants to occur, it is not sufficient. So, for example, left and right limb buds are developmental modules. The experimental manipulation of one need not perturb the other. But they are not morphological modules in Lewontin's sense; they do not vary independently of one another in ways that yield selectable variation. Even so, provided that under most circumstances the modular outputs of development are potential modular inputs for natural selection, we can treat developmental modularity as a surrogate for quasi-autonomy, and, perhaps, vice versa.

We think Günter Wagner is right to introduce a notion of modularity that explicitly combines both evolutionary and developmental notions. Wagner's program would forge a link between a developmental mosaic and the local morphospaces that we discussed in 4.4 through 4.6. In representing a clade's morphospace, the dimensions to represent are the dimensions of potential variation. In Wagner's view,[9] to understand evolution in a lineage, we have to identify those traits that are the real, objective building blocks out of which organisms are built. These are characterized both developmentally and evolutionarily; Robert Brandon (1999) has developed a similar view. Wagner's building blocks exist within a species or lineage in a number of variants. These variants, like the alleles of a gene, can potentially replace one another. They have distinctive effects on fitness. It makes sense to ask of one of these modules: what is it for? These blocks out of which phenotypes are built are also, of course, developmentally distinctive. Their development chiefly depends on a small chunk of an organism's genome, though within that chunk the effect of one gene depends on the effects of others, and each gene has multiple effects. The idea is captured schematically in figure 5.5, where each of two small sets of genes distinctively influence a particular trait complex, each of which in turn makes a distinctive contribution to the functioning of the organism.

Notice that for Wagner, modularized traits have both evolutionary and developmental characteristics: these traits have identifiable

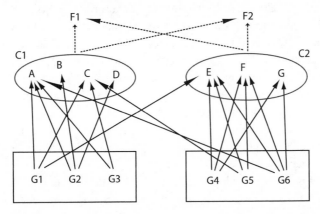

FIGURE 5.5. The genotype-phenotype map. C1 and C2 are character complexes, each having a primary function (F1 and F2). This is an example of modular gene expression, because the pleiotropic effects of G1–G3 are primarily on C1 and those of G4–G6 are primarily on C2. But note that modularity is a matter of degree. After Wagner and Altenberg (1996).

functions; they have coherent selective histories. They make a distinctive, identifiable contribution to the fitness of the organism of which they are an identifiable aspect. And they have distinctive developmental histories. The evolutionary aspect of Wagner's characterization is important, for it's the basis of his attempt to model the evolution of developmental modularity. The mosaic organization of development influences developmental possibility. But it has an evolutionary history itself, and it is not set in stone; evolutionary plasticity itself evolved and evolves.

Günter Wagner and Lee Altenberg argue that development becomes more modular when we have stabilizing selection on most of the phenotype but directional selection on one aspect of it.[10] In a finch lineage under sustained directional selection on beak shape, but under stabilizing selection on the rest of its phenotype, heritable variations that sever the developmental connections between beak shape and other aspects of phenotype will be favored. Such opposing selection pressures select for any gene changes (for so-called modifier genes) that reduce the interactions between the two sets of genes and their associated traits. Likewise, modules that depend in development on the interaction of many genes can be split apart when their multiple phenotypic outcomes have opposing fitness values. Hence, over the course of evolutionary history, larger modules tend to become composed of smaller submodules offering more scope for diversification and specialization. In some of his more recent work, Wagner suggests that modularity can increase as a side effect of stabilizing selection. Stabilizing selection selects for making the development of the trait as reliable as possible. In turn, that selects for making the development of the trait insensitive to noise—noise in the world, and noise in the genome. The greater the number of genetic inputs to the development of a trait, the more opportunities there are for its development to be perturbed by genetic and developmental noise. So stabilizing selection can decrease sensitivity to genetic inputs (Wagner et al. 2005).

These models are clearly pretty conjectural, but they do show that the developmental ideas can be put into a plausible evolutionary context. That said, the developmental and evolutionary notion of modularity needs some adjusting in the light of developmental plasticity and its importance. In discussing developmental interaction, Wimsatt, Wagner, and Kauffman all make what we might call the "mutational assumption." The mutational assumption is that when a novel gene or gene combination occurs, its effects on phenotype (if any) are undirected with respect to fitness. If the effect is small, it is no more likely to be advantageous than disadvantageous; if the effect is large, it is likely to be disadvantageous. Wimsatt's generative entrenchment model, likewise, assumes that if X and Y are two traits interconnected in development, and if the development of X changes in ways relevant to Y's development, those effects, like the effects of mutation, are undirected with respect to fitness. We have seen in 5.2 that this is too simple. Organs that develop together are linked in ways that make a response by one to a novel development in the other likely to be adaptive. Evolvability is not maintained and could not be maintained by imposing a developmental quarantine on each organ or organ system in the developing organism; that would lead straight to the evolutionary coordination problem we noted in 5.2. So our view of the developmental interconnections that limit evolutionary change must be more nuanced. Some developmental mechanisms that link the development of two systems enhance evolutionary plasticity rather than limit it. So developmental autonomy is not quite the right conception of modularity; quasi-independence itself depends on some combination of autonomy with adaptive plasticity.

5.5 DEVELOPMENTAL BIODIVERSITY

Lineages differ in evolutionary plasticity. It is clear that plasticity is influenced by population-level properties, and that these vary across populations. Small populations with little variation are more vulnerable than large populations with more variation. Presumably, there are also differences in the developmental systems of individual organisms that result in plasticity differences across lineages, though the empirical issues are less tractable in this case. We take it that the evo-devo literature has clearly demonstrated that biases in the supply of variation are capable of influencing the trajectory of evolution. But we do not think there are clear demonstrations about the extent, nature, or relative frequency of such biases, though there are certainly plausible suggestions about specific cases (Arthur 2001). Even when comparing living taxa, it's hard to operationalize the crucial concepts. For example, Andrew Yang makes a brave attempt to test the idea that modularity is connected to evolvability by comparing the disparity of hemi- and holometabolous insects. Holometabolous insects experience full developmental transformation over their life cycle, while the nymphs of hemimetabolous ones resemble the adults. As it happens, holometabolous insects are much more diverse, but it is far from clear that their development is more modular in the relatively crisp senses that Wagner identifies (see Yang 2001). So a difficult conceptual and empirical task remains: that of identifying a clear and empirically tractable notion of developmental modularity, and making some progress on identifying the extent to which development is modular. Developmental biology has been based on a small group of model organisms, and these have typically been chosen for their empirical tractability rather than because their development is known to be typical of the clades they represent (see Jenner and Wills 2007). Moreover, the relationship between the system of genetic switches identified as important in early development, and morphological traits, remains to be established. So these conceptual and empirical tasks are daunting.

We might suppose that the attempt to directly establish differences in plasticity across different lineages using developmental biology could be supplemented by paleobiological information. There are, after all, enormous differences in species richness and ecological penetration among the living bilaterian phyla. For example, only a few have ever established in fully terrestrial environments (mostly arthropods, chordates, annelids, mollusks). What do these differences show? As Rudy Raff notes, there are no centaurs. No six-limbed vertebrates have ever evolved from four-limbed ancestors. Is this evidence of the developmental impossibility of centaurs? How can we tell from the fact that the elements in a trait cluster *did not* diverge independently of one another, that they *could not* evolve independently of one another? The inferential problem is difficult because we must take into account a potential distinction between a capacity and the expression of that capacity. Even if a clade is evolutionarily labile, and is labile in virtue of ancient features of its developmental biology (rather than recent innovation), that lability need not leave a signature in the pattern of morphological evolution. Evolutionary radiations and morphological innovations require a cooperative environment—the right selective histories—not just a cooperative supply of variation.

Consider the fact that so few clades have established on land. Perhaps we can assume that the deep history of arrow worms or cephalopods must at some stage have favored amphibious forms, were they to have been available to selection. Octopuses that live in tidal pools, for example, are often forced to travel from one to another over wet rock, so surely (the thought goes) there would have been situations in which the ones that could travel farther or survive longer out of their holes would have been favored. Thus we can infer from the fact that no terrestrial octopus has ever swung from the trees of a rainforest, plucking monkeys from their perches, that this clade lacks the potential for transition to a terrestrial world. We think it is indeed plausible that some differences in plasticity have contributed to cephalopods resolutely seafaring ways (though we do not see how we could prove this to a skeptic). But it is hard to see how to assess the relative importance of cephalopod developmental systems and external factors; perhaps the lineages that have radiated on the land have done so mostly because they happened to have the right morphology at the times at which ecological windows of opportunity opened. Thus while paleobiological information about the striking differences in ecological penetration between different clades is intriguing and suggestive, that information by itself does not demonstrate differences in plasticity.

Time once again to connect the threads of this chapter to the previous discussion. The last few chapters have all focused on the role and limits of species and their pattern of relationships as a summary and surrogate for biodiversity. The history of life (or, at least, the history of the macrobes) is captured to a considerable extent by species lineages as they branch, diverge, and extinguish through time. Species richness in its phylogenetic context is a slice through that history, a slice that in part reflects previous history, in part predicts future developments. But only in part. What we need to add to this basic information, though, will depend on our particular interests. As we noted at the end of the last chapter, if our interests are in the mechanisms that made the present, we will need to explicitly add in information about phenotypes and phenotype change. We may need to add developmental information, too, if these patterns are influenced by biases in the supply of phenotype variation. If we want to project our understanding of the biota as it is now into the future, likewise, we will need information about both short-term plasticity (for example, information about standing variation) and about medium- and long-term influences on the supply of variation to ecology and evolution. The spread of phenotypes in biota is important. So too is the way those phenotypes are parceled out into distinct lineages. And so too are the developmental mechanisms that supply or withhold further phenotypes to that spread. Assessing the evolutionary potential of a taxon is not an easy task. Species richness is probably, in many contexts, a good surrogate for both morphological disparity and evolutionary plasticity. The more we learn about the evolution of development, the more we will learn about the strength and scope of this correlation. As we have already noted though, we can only show that species richness is a good multipurpose measure of biodiversity if we can independently represent both disparity and plasticity. Despite the serious empirical and conceptual challenges it faces, the idea of local, clade-specific morphospaces grounded by conserved and shared developmental systems seems to be the most promising way of representing both disparity and plasticity.

6 *Explorations in Ecospace*

So far in this book our main focus has been on evolutionary dimensions of diversity, and on the causes and consequences of that diversity. A central theme has been the relationship between species richness and other dimensions of biodiversity, and the extent to which the biodiversity of a system is captured by information about the identity, demography, and evolutionary relationships of the species in the system. While species richness does not determine these other dimensions, and may not always be a good surrogate for them, there are important causal and theoretical links between species richness, morphological disparity, and plasticity. In the previous chapter we looked at recent work on evolution and development. We discussed the relationships between species, evolutionary morphospace, and the supply of variation to evolution. In this chapter our focus changes to ecology. We explore ecological notions of diversity and the relationship between ecological and evolutionary systems.

We begin with a familiar question: What more might we want to know about the biodiversity of an ecological system over and above its species composition and facts about variation and plasticity within those species? Ecology is an enormous and complex field in its own right, so we will pursue this general theme through a specific example. This chapter focuses on local ecological communities, and on whether local communities are structured, organized systems; that is, systems whose organization has important effects on the identity and abundance of the local biota. In analyzing the idea that communities are indeed structured systems, we will consider the claim that communities control their own membership and the claim that they have biologically important collective properties. If these ideas are vindicated, we do need more than species information. We need information about organization and variation in that organization from community to community. In this chapter's "units-and-differences" framework, we ask whether local ecological communities are themselves units, and, if so, what are the relevant similarities and differences among them.

Ecologists study (among much else) the interaction among populations, and among those populations and their environments. Their aim is to understand the distribution and abundance of organisms. They study the processes that determine distribution and abundance at many spatial and temporal scales. But perhaps most attention has been focused on local communities, on the avifauna of a particular wood, or on the invertebrates on a particular beach. As Robert Ricklefs has put it, much of ecology has been organized around a model of "local determinism" (2004). On these models, the abundance and composition of local communities is essentially controlled by the causal characteristics of that community itself. Thus, much ecology has been local community ecology, and we shall follow that lead. We do so with some reluctance, for one of the most interesting recent developments in ecology is a shift away from local determinism to macroecological models (for a recent review of the many different proposals about the natural units of ecology, see Jax 2006). On these macroecological models, the profile of a local community (the species present and their abundance) is driven mostly by the characteristics of regional biotas. A forest patch in Ecuador is more species rich than a similar-sized patch in England because the Ecuadorian regional biota is immensely more rich; it is not the characteristics of the patch itself that primarily explain this difference.[1] We return to the relationship between local and regional structure briefly in the final section, but our focus on local determinism implies that this chapter should very much be thought of as a preliminary study of ecological diversity.

Local determinism could be true in two ways. The identity and abundance of the organisms in a local patch might be controlled by the abiotic environmental features of the patch: rainfall, temperature, soil profiles, wind exposure, and the like. Alternatively, it might be controlled by interactions between the organisms present, interactions that favor some potential residents and exclude others. (Obviously, mixed models are possible.) Thus one important issue in ecology is whether the distribution and abundance of organisms is in part explained by characteristics of ecological systems themselves. Are local communities *organized systems* that make available space for some populations and exclude others, thus regulating their own membership? Ecologists began with the view that local systems were organized systems in this sense. Charles Elton's theory of the niche, the first theory of ecological niches, took

niches to be the ecological equivalents of economic roles in human social systems. The organization of a particular community made certain
ways of life available within it, but not others (Griesemer 1992; Worster
1994). But there were early dissenters who argued that distribution and
abundance is largely explained individualistically (Gleason 1926). Different species have different tolerances to physical conditions, different
resource requirements, and different levels of vulnerability to biological threats or physical disturbances. The distribution of organisms is
largely explained by these species' independent responses to variation
in the environment, especially the physical environment. If individualist models of ecology are vindicated, information about the presence
and abundance of species captures the ecologically relevant biodiversity
of biological systems.

There is clearly some truth in the individualist idea; the distributions, abundances, and evolutionary trajectories of particular species are
profoundly influenced by the physical features of their environments.
Arid, nutrient-poor Australian environments have a very different biota from that of the (barely) temperate rainforests of the west coast of
New Zealand's South Island. To some extent, biological variation across
space and time is a response to physical variation across space and time.
In this sense, the investigation of ecological diversity—biotic variation
across habitats—is a calibration of the physical parameters that affect
the distribution of species. In a well-known example of work of this
kind, Robert Whittaker has argued that much of the variation in species
composition across different habitats can be explained by just mean annual temperature and rainfall (1975, 167).

To the extent that the distribution and success of organisms can be
explained by their autonomous response to such physical variables, the
individualist program in ecology will be vindicated. We will not need to
appeal to ecological systems, to the structure and organization of communities, to explain those facts. Rather, distribution, abundance, and
fate are explained by the interactions between species' evolved phenotypes and their environment. Suppose, for example, that the replacement of rimu-kahikatea forest by southern beech as one goes south on
New Zealand's west coast can be explained by these species' differential
responses to temperature, rainfall, wind, and soil. If so, we would not
need to appeal to features of rainforest community organization—to
features of the ecological system—to explain these species' distributions and abundances. Individualists recognize that many organisms
need biologically made resources, but their bet is that the biological
tolerances of local populations are, for the most part, quite coarse. Of
course there are specialists. Glossy black cockatoos (*Calyptorhynchus*

lathami) require mature casuarina trees, and some caterpillars will lay their eggs on only one species of plant. But while organisms depend for resources on their biological as well as their physical environment, in the individualist view they do not typically depend on a specific array of interacting populations. Species do not really care who their neighbors are.

In assessing the plausibility of this individualist line of thought in ecology, it is important to distinguish between a phenomenological and a causal view of local communities and kinds of communities. There is no doubt that local communities—these assemblages of plants and animals found in association on distinctive habitat patches—are part of the descriptive phenomenology of ecology. Moreover, there is a reasonably natural and predictive taxonomy of habitat patches. For example, when we find out that there are tidal mudflats near Moruya (on the New South Wales coast of Australia) we have a fair idea of the plants and animals we can expect to see: mangroves, samphire, a suite of distinctive birds (herons, waders, and the like) and so on. Coastal wetlands on the east coast of New South Wales vary one from another, but nonetheless they support a broadly similar range of species. These similarities allow field guides and similar tools to distinguish among woodland and closed forest, wetlands, grasslands, coastal heathlands and sand dunes, and arid and semiarid areas. Flora and fauna differ in characteristic and fairly repeatable ways that are captured by these descriptions. We can identify certain types of community by statistical patterns of association among species; woodlands and wetlands have different inhabitants. In identifying community types this way, we say nothing about the processes that produce these identifiable and repeated associations. So communities are important to ecology in this minimal sense; local patches have stable natural histories that typically do not vary dramatically from year to year. And a given local patch will resemble some other local patches well enough for there to be useful taxonomies of habitat types. The modest view of these local systems, then, is that they have reasonably stable natural history profiles, and that fact enables us to make some reasonably reliable qualitative predictions about their overall biological composition. Different phenomenological communities may just reflect differences in interactions between species and their physical environments. But they would still be useful surrogates: allowing inference from community type to species composition.

Thus the phenomenology of local communities is important because it reveals what we want to explain and protect. It is important for a second reason. Differences in the natural histories of these phenomenological communities are symptoms of important ecological processes. For example, in Wellington there is a so-called mainland island, the Karori Reserve. This is a chunk of remnant bushland that survived around Wellington's former water reservoir. It has now been enclosed with predator-proof fences, and it is the site of a major effort to extirpate exotic mammals and weeds. The local Web site has reported the striking results of these changes: endangered animals, like the little spotted kiwi (*Apteryx owenii*), have been successfully reintroduced to this area. And birds like the tui (*Prosthemadera novaeseelandiae*)—once only just hanging on in the city—have rebounded. The differences between this reserve and other areas of local bushland (and from its former self) are only too obvious. It is full of native plants and animals. They are infested with possums, hedgehogs, rats, feral cats, and an assortment of weeds. At the Wilton's Bush Reserve, only a kilometer or two from Karori, the difference is obvious even in the course of a short walk. Wilton's Bush is quite rich in native vegetation, but the forest is almost silent.

Even if local communities have no causally salient properties that drive ecological processes, it is no surprise that ecological and conservation biology journals are full of descriptions of local communities. The differences between them reveal ecological processes in action: competition, predation, response to physical disturbance, and invasion. The local biology students learn their survey techniques at Karori, looking for mouse droppings and counting birdcalls on transects. Phenomenological descriptions of local communities help set and test an explanatory agenda for ecology and, often, for conservation biology as well. The difference between the Karori Reserve and Wilton's Bush is not the result of deliberate environmental manipulation to test for the ecological consequences of introduced predators; Wilton's Bush is not a deliberate control plot. But when conservation biology meets ecology, a standard experimental probe is to match "no-intervention" communities with "intervention" communities. Thus the devastating effects of foxes on small to medium size marsupials has been established by contrasting communities in which fox numbers are suppressed by baiting, with communities without fox control, and surveying the marsupial fauna of interest (Kinnear et al. 2002).

The effects of crucial biological processes are often revealed in this way, by comparing communities that differ (as far as we know) only in one important respect. For example, one important debate in contemporary ecology is about contingency. Contingent systems are sensitive to unpredictable events, and hence their future trajectories are unpredictable. One form of contingency is "path dependence." A community's future is path dependent if (for example) the order in which migrants arrive makes a major difference to the community that is ultimately assembled. If order effects were important, we would not be able to predict the future trajectories of island communities because their local ecology would depend on the accidents of arrival order. If species A and B were to arrive together, B would exclude A. But if A arrives first, it has a good chance of preventing B from establishing. It clearly matters whether path dependence is ecologically important, and the best way of testing for path dependence is by comparing communities with similar early histories to determine whether their futures are similar when they differ only or mostly in the order in which colonists arrive. The Krakatau islands have provided the opportunity to make just these comparisons, as the different remnants of Krakatau allow comparisons of different islands at the same time, and of the same island at different times (for further eruptions have turned the assembly clock back to zero).[2]

So one way of thinking about ecological diversity is in terms of a phenomenological ecospace. The dimensions of that space include salient measures of the physical environment. For terrestrial communities, these are rainfall, temperature, soil structure, and the like. Such an ecospace will have biological dimensions, too, specifying the presence and abundance of the dominant vegetation (by species or by functional group), and likewise for other trophic layers. Two wetlands in southeast Australia will end up near neighbors in such a space, in virtue of their physically similar substrates and the presence of similar organisms in similar numbers. The dimensions, then, are the dimensions of descriptive ecology: physical environmental variables, vegetation cover, and the animals living in and on the vegetation. As with morphospace, though, a total ecospace is of high and somewhat arbitrary dimensionality. Would we have a dimension for every duck in the regional biota? A dimension for every soil element, or just an aggregate measure of fertility? Ecologists will typically be interested in comparing communities with respect to just a few dimensions, and those few will depend on the purposes of the comparison. If we are interested in the impact of foxes on small marsupials, the most crucial dimensions will be fox abundance and small marsupial abundance, though if we think other factors might exacerbate or mitigate the effects of foxes, we will have to include these too (for example, density of cover, other predators). For fox-baiting studies, the boundaries of the community are defined by the boundaries of fox-baiting, for we are interested in the effects of baiting in that region, and that is true whether or not the limits of baiting coincide with an overt phenomenological change in the local ecology. For other purposes, we would represent the same habitats quite differently. A botanist interested in the causes of eucalypt dieback would choose different dimensions of comparison.

Even if the individualists are right about local communities, a good descriptive taxonomy of local communities would be a good tool for both conservation biology and ecology. It would deliver an easy to use surrogate for alpha and beta species richness,[3] and a valuable probe for assessing the impact of ecological processes. But there is a more ambitious project, one that takes local communities themselves—as distinct from the populations that comprise them—to have causally salient properties. The crucial question here concerns the extent to which local communities are functionally organized systems. Consider, for example, Black Mountain, a eucalypt woodland community in Canberra, and one of Sterelny's local patches. Is this an organized biological system? Not if individualism is right. If the Black Mountain community is an assembly of populations whose sizes and prospects of persistence are largely independent of one another, if its components have impacts on their environment that are mostly independent of their neighbors, and if it is an assembly of populations with varying ranges that somewhat overlap, then the Black Mountain community would *merely* be part of the descriptive phenomenology of biology. It would be a "unit" in something like the sense that a genus of duck species is a unit, rather than the sense in which a species is a unit. Identifying Black Mountain as a eucalypt woodland on the southwestern slopes of New South Wales would give conservation biologists a good idea of its alpha diversity. The differences between it and otherwise similar phenomenological communities calibrate the power of ecological mechanisms. But there would be questions it makes no sense to ask. The community would have no objective bound in space or time, and nor would the community as a whole have explanatorily salient features. So from the fact that such communities and community types can be characterized phenomenologically, it by no means follows that communities have autonomous, biologically important properties; it by no means follows that they have organizational or structural properties that help explain the distribution and abundance of organisms.

In brief, the individualist idea has led to a very serious debate about the extent to which communities are organized systems, whether community structure filters the species present in a local patch, excluding some and admitting others, and whether that same structure determines (or strongly constrains) the abundance of those populations that are present. In the language of ecological theory, there has been a debate about the extent to which communities are structured by "assembly rules," rules that specify those species that can, and those that cannot, co-occur with one another in local communities.[4] It is possible that, say, a local population of banksias and another of eucalypts on Black Mountain are associated spatially only because both populations happen to tolerate the temperature, soils, and rainfall characteristic of this location. If each of the Black Mountain populations is more or less indifferent to the presence of others, then this assemblage is merely a "community of indifference."[5]

Communities of indifference are merely phenomenological communities; populations within them are spatially associated only because they happen to tolerate similar physical conditions. Such "communities" are not organized, structured systems. On this view, there will be a more or less deterministic explanation of why particular species are represented on Black Mountain—soils, rainfall, and temperature make it hospitable to some members of the regional species pool but not others. But these explanations will be relatively independent of one another. An explanation of the composition of the community is no more than the sum of the explanations of the presence of each member of the community. Communities of indifference have no causally salient organization. Yet, if community regulation is important, so that membership and abundance is filtered by the structure of the community itself, then communities are not just part of the phenomenology of biology. The populations present and interacting in a particular local habitat constrain the range of potential members of that community. The current status of this idea is the focus of 6.3.

There is a second challenge to the idea that ensembles in a local patch are just communities of indifference. Communities are real, causally important ecological systems if they have emergent or ensemble properties; if, for example, a forest dominated by pines has properties that are not just an extrapolation of the properties of individual pine trees. This idea is controversial, and we will return to it in 6.4. Within ecological theory, the idea that communities have ensemble properties has been explored in many ways. We shall do it by considering the diversity-stability hypothesis, the idea that more diverse communities are more stable. According to this hypothesis, diverse communities are less perturbed by external disturbance, and they return to a predisturbed condition more readily than less diverse ones. Diversity in this context is a property of the community itself, and so, in some version of the stability-diversity hypothesis, is stability. In diverse communities the overall productivity suffers less in (for example) drought, but individual populations may fluctuate as profoundly as those in less diverse communities. The idea here is that communities themselves (as distinct from the organisms and groups that compose them) have causally salient properties.

6.2 COMMUNITIES, ECOSYSTEMS, AND ECOSYSTEM FUNCTIONS

We think that communities are causally important, but that particular communities vary in the causally salient properties they have and the degree to which those properties are causally salient. We begin, though, with a conceptual prologue: function and functional organization in ecology. Function in ecology is not like function in evolutionary biology or functional morphology. In those fields, functions derive from selective history (Wright 1973; Millikan 1989; Godfrey-Smith 1994). The ponyfish has a light-emitting organ, and the function of the light the organ generates is to prevent the ponyfish from being visible from below, silhouetted darkly against a lighter background. In matching the illumination radiating down from above, the ponyfish is concealed from predators. The ponyfish shines to be invisible. In making this claim about the function of the light-emitting organ, we make a claim about selective history. Ancestral ponyfish with such organs survived better than those without them (or with less well-tuned organs) because they were less often seen from below (Williams 1997).

It is not likely that we can explain functional roles in local communities in a parallel way. In the early history of ecology, the idea that communities were like organisms was taken quite seriously. Frederic Clements (1936) thought of communities as systems in a very rich sense, as akin to superorganisms. He based this on his view of ecological succession. Succession organized plant communities in a robust way, so that even after very severe disturbance, a homeostatically preserved equilibrium, the climax community, would be rebuilt (Cooper 2003). But no one would now defend a view of functional organization of communities modeled on the functional organization of organisms. Not only are organisms much more tightly integrated and bounded than the typical community, but also, as a rule, local assemblages do not have selective histories. They are not part of lineages. Communities are not elements of a population of competing communities, and they do not have daughter communities that resemble their parents. If a selective history is necessary for communities to have organization or structure, then most assemblages of populations are not ecological systems.

However, as Robert Cummins has shown, there is an alternative view of function and organization. A part of a system has a Cummins-function when its activity makes a distinctive, stable contribution to the operation of the system as a whole (Cummins 1973; Godfrey-Smith 1993, 1994). Thus in many Australian woodlands, eucalypt litter has the Cummins-function of making fire more likely. This is a stable, regular contribution of this component of a woodland system to the overall behavior of that system. So local communities may be functionally organized, structured systems because their components have Cummins-functions. For example, there has been important work in ecology, beginning with Robert Paine (1966), on the role of keystone predators in maintaining diversity. They do so by limiting populations that would otherwise out-compete others (for a review, see Power et al. 1996). Keystone species have Cummins-functions; starfish are not selected to maintain diversity by eating mussels, nor has there been between-community selection for mechanisms that maintain diversity. But within that community, this is a stable effect of this particular population (see Box 6.1). Community ecologists often analyze communities in terms of guilds or functional groups, which are components of a community intermediate between a community as a whole and a local population (Naeem 1998). They are sets of populations playing specific roles within a community: browsing, pollination, or seed dispersal. Such guilds and functional groups are identified by their Cummins-function (Blondel 2003).

Ecosystem ecology, in particular, has been centrally concerned with identifying and explaining Cummins-functions. There has been a

---

BOX 6.1: Keystones and Dominants
A keystone species is one "whose impact on their community or ecosystem is large, and disproportionately large relative to their abundance" (Power et al. 1996, 609). Dominants are species that are very abundant in an ecosystem and that "play a major role in controlling the rates and directions of many community and ecosystem processes" (Power et al. 1996, 609). The number, power, and the location of interactions are all of importance to a population's ecological impact, as Jordán and Scheuring (2002) note (as in fig. 6.1).

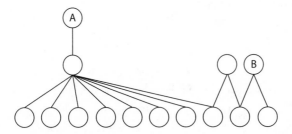

FIGURE 6.1. This hypothetical food web shows that the number of links may be a misleading measure of the positional importance of species. Species A has a single link, but it is in a key position, while species B has two links, but its effects spread less easily to other members of the community. After Jordán and Scheuring (2002).

historic divide between community ecology, aiming to explain the identity and abundance of local species populations, and ecosystem ecology, aiming to explain the flow of material and energy through the local system (Golley 1993). For example, ecosystem ecologists study the flow of crucial nutrients like phosphorous and nitrogen from the soil into organisms and back into the soil. The organisms responsible for these flows—the detrivores that consume litter and make soils—are performing Cummins-functions; they make a stable, repeatable contribution to the behavior of the system as a whole. Despite this historic divide between ecosystem and community, it will not be pivotal to our discussion. John Odling-Smee and his co-workers have argued that the distinction between community ecology and ecosystem ecology is eroded once community ecologists recognize the niche-constructing role of organisms and populations (Odling-Smee et al. 2003). Organisms do not just eat, breed, and die. They reorganize their environment. Hence, an explanation of the presence, abundance, and activities of local populations will also explain the biotically caused flow of materials and energies through that local system. Once the role of organisms in niche construction is recognized, the distinction between community ecology (focusing on the distribution and size of populations) and ecosystem ecology (focusing on the flow of matter and energy through a habitat) becomes much less sharp.

In terms of this framework, then, phenomenological communities are organized systems only if they are stable, bounded, and with enduring global features of biological importance to which particular components make a regular contribution. Arguably, they are organized systems in this sense if they are regulated, that is, constrained in membership and numbers by their Cummins-functional organization, or if they have causally important emergent properties.

6.3 INDIVIDUALISM AND COMMUNITY REGULATION

Local populations do not live independently of one another. Species depend on the local biology; there can be no echidnas without ants. But individualists think that species have broad-banded biological conditions of existence. Most particularly, competitive interactions do not determine community make-up; species are not typically excluded by other species with similar resource-use profiles. So they are skeptical about the predictive importance of an important organizing idea in ecology, the principle of competitive exclusion. The principle itself states that species with the same resource requirements cannot indefinitely co-occur; one will be competitively superior and drive the other to extinction. Generalizing this, species with similar requirements will have strong competitive interactions, and will tend to exclude one another. These results are based more on models than on observations of natural systems, and many ecologists doubt that real habitats are sufficiently uniform and stable to reach the equilibrium at which exclusion takes place (for reviews, see Kingsland 1985 and Cooper 2003). Suites of parrots, of honeyeaters, and of insectivores coexist on Black Mountain and they do so (according to this line of thought) because real habitats are heterogeneous; many populations extend over patches that contain relevant environmental variation. So one species of thornbill does not exclude the others. Moreover, they are fluctuating. They are not filtered by competitive exclusion, because the world intervenes before local assemblages reach their theoretical equilibriums. On this individualist view, phenomenological communities are typically associations of overlapping populations. Such phenomenological communities do not have determinate boundaries. Moreover, though these populations are not fully independent of one another, they interact weakly. Populations do not regulate one another, nor do they impose hard-to-penetrate filters on community membership.

How plausible is this conception of communities? In particular, is it consistent with the readily observed, qualitative stability of local ensembles of populations? As Greg Cooper discusses at some length, there is a line of thought in ecological theory that infers regulation from stability.[6] Stable ensembles, the thought goes, are internally organized through competitive interactions. The stability that makes field guides possible cannot be explained by abiotic factors. Their impact is too variable. Rainfall, for example, varies dramatically from season to season, and so too does the incidence of fire. The Black Mountain biota does not inhabit a physically invariant landscape. Yet the Black Mountain community is roughly stable in both composition and abundance. That fact is best explained by the hypothesis that communities are regulated by "density-dependent" biotic interactions. The size of some given population—say, superb fairy wrens on Black Mountain—will fluctuate within bounds only if the factors that limit the fairy wren population become *more intense* as the population rises, and *less intense* as it falls. An obvious candidate for such a factor is competition between the wrens for limited resources. The more wrens, the harder such limits bite. Competition is bound to get more intense as population size increases, and less intense as it dips.

In brief, stability is the result of a "balance of nature," a balance deriving from the internal regulation of communities. The qualitative stability of natural and artificial ecosystems shows the importance of density-dependent factors. If the forces that affect a local population act independently of its size, it would be an amazing coincidence if abundance did not change over time. Very slight tendencies to increase or decrease result in crashes or plagues. If populations persist, something must damp down such fluctuations. Yet abiotic factors are not sensitive to population size. The impact of flood, fire, or drought—and external disturbances more generally—is not sensitive to the size of the populations on which they impact. An oil spill will destroy a seabird rookery without regard to the number of birds present. We can infer from the qualitative stability of communities that they are networks of biological interaction that filter membership and that constrain the demography of their members.[7]

Cooper is rightly skeptical of this whole class of arguments; they depend on a crucial ambiguity (Cooper 2003). There is an undemanding sense of "stable," where it means something like "the persistence of community membership." On this reading, communities are indeed typically stable, as is Black Mountain, whose species composition is similar year by year. But while most communities most of the time show a fair degree of persistence of community membership, that does not establish the existence of internal regulating mechanisms. Over shorter periods, other mechanisms can explain persistence. The crucial point is that communities are often demographically open. Thus a local population may persist by recruiting from neighboring communities. The stability of demographically open communities can be the result of such metapopulation dynamics. If, for example, echidna populations vary independently of one another in a cluster of adjacent communities, a population fluctuating toward extinction can be rescued by migration from a neighboring community whose numbers happen to be surging. Migration between communities can protect unregulated communities from random walking to extinction. The effects of density-dependent internal regulation can be coarsely mimicked by a metapopulation of unregulated communities, provided that metapopulation is spread over a heterogeneous landscape and provided that migration from one population to another is possible.

Populations without density dependence can persist for many generations. Even if competition, predation, and other density-dependent ecological mechanisms are not important, so long as the trajectory of populations within a cluster are independent of one another, the stability of the metapopulation ensemble will be greater than the stability of a typical population within the ensemble (Baguette 2004; Hanski 2004; Murdoch 1994). We do not know the extent to which metapopulation dynamics explain the evident stability on which field guides depend. But the existence of this mechanism means that we cannot assume that persisting communities are internally regulated. Moreover, we know there are qualitatively stable associations that can hardly be the result of strong interactions between the local residents of a community. There are field guides to estuaries and other habitats where many of the birds are migrants; they are winter residents. Many of the waders found on Australasian tidal mudflats breed in the far north. And while banding studies suggest that these birds are faithful to their breeding zones, there is no reason to believe that the same birds—the godwits, the knots, the turnstones, and curlews—return year after year to Foxton estuary on the east coast, north of Wellington, or to Miranda, south and east of Auckland.[8] The stability of these associations is presumably explained by a stability in the flow of resources through these systems.

A more demanding notion defines stability not just in terms of community membership but also of population size. If communities are at true equilibrium, with population sizes varying only in minor ways around a mean to which they typically return, then they must indeed be regulated. But so understood, there is no reason to believe communities are typically stable. In short, if there is evidence of limited variation around a mean in the population sizes of components of the community then we do indeed have evidence of equilibrating mechanisms. But it remains to be shown that local assemblages are typically characterized by limited movement around a mean.

Time to take stock. It is an obvious truth of ecology that local assemblages are fairly stable over short periods of time.[9] Farming would be an impossible activity if that were false. But while this fact is suggestive, in itself it is not sufficient to show that local assemblages are typically ensembles of strongly interacting and thereby stabilized populations. If stability just consists in the persistence of community membership, such persistence may be explained by metapopulation effects. The observed phenomenology of stability does not show individualism to be mistaken. What, though, of emergent properties? Perhaps communities have causally important properties that equilibrate features of the community as a whole: diversity, productivity, or ecosystem services (for example, the flow of crucial nutrients like nitrogen and phosphorous from organisms to soils and back). This is an important idea to which we turn in the next section.

6.4 THE EMERGENT PROPERTY HYPOTHESIS

In this section we discuss the idea that local communities have causally salient properties by exploring a family of famous hypotheses that link the diversity of a community to its stability. Communities (the thought goes) have emergent properties. An ensemble has emergent properties if it has features that are not simple reflections of the properties of its parts. The notion of an emergent property is not inherently mysterious or spooky. No one doubts that organisms have emergent properties. Organisms are built from cells (plus some of their products), and it is quite clear that the fitness of an organism (for example) is an emergent property of the cell ensemble in its environment. There is no way that the fitness of a particular eucalypt is a simple reflection of the cells, cell population, and their products out of which the eucalypt is built. The fact that organisms are composed of cells was an epochal biological discovery; there is no understanding how organisms work without understanding how cells work. But it does not follow that we can understand how organisms work *just* by understanding how cells work. The compositional structure of ensembles constrains their behavior, and hence is crucial for understanding system-level behavior. However, it may not be true that understanding the parts from which an ensemble is built suffices for understanding the ensemble.

Thus, in discussing the relationship between an ensemble and its constituent elements, philosophers distinguish between a supervenience claim and an explanatory claim. The supervenience claim is that there can be no change in system-level properties without change in the properties of the parts; the Karori Reserve community cannot become better buffered against invasion by exotic weeds unless there is some change among the particular populations that make up the community. The explanatory claim is that all important system-level behavior can be explained, and can only be explained, by explaining the behavior of the parts. For such cases as the relation between organismal properties and those of cells, or the relationship between communities and the populations within them, the supervenience claim is uncontroversial. But the explanatory claim is controversial. (See Jackson and Pettit 1992; Sterelny 1996; for the claim that emergent properties are inescapably spooky, and science should never posit their existence, see Rosenberg 2006.)

The crucial idea of the emergent property hypothesis is that these emergent properties are causally important; they drive ecological processes. The diversity-stability hypothesis is one attempt to show this. The thought that diversity adds stability to a community has enormous intuitive plausibility. Diversity adds redundancy, and hence allows that community to survive fluctuations in the fortunes of its members. If only one population on Black Mountain pollinates the gum tree, *Eucalyptus rossii*, and if it were to suffer a serious decline, *rossii* would be unable to recruit new plants into its population. However, if there were a suite of eucalyptus pollinators, a fluctuation in one population would not

ramify through the community as a whole. Redundancy buffers disturbance, and diversity adds redundancy.

This appealing picture seemed to be undermined by the theoretical
work of Robert May (1973). His models showed that more diverse communities were less stable, not more stable. The last decade or so has seen
a revival of the diversity-stability hypothesis and its close relative, the
idea that more diverse communities are more productive. Interestingly,
both May and his critics identify diversity with species richness. From
the perspective of this book, it is clear that his account of diversity involves an important simplifying assumption. Rather than taking up this
simplifying assumption, the critics have sidestepped May's result, taking
issue with May's account of stability. May took the diversity-stability
hypothesis to be a hypothesis about *population size*; in more diverse communities, the populations of the component species are more stable.
However, David Tilman and others have argued that *community-level
properties* are more stable in more diverse communities. Tilman proposed changing the focus from community composition to ecosystem
processes, and to the connections between those processes and community composition. In particular, Tilman argued that the biomass of more
diverse communities is more stable[10] than that of less diverse ones. For
somewhat similar reasons, a diversity-productivity relationship looks
plausible. Habitats are heterogeneous in space as well as time. A habitat patch will exhibit small-scale variation in its physical and biological
characteristics, and so (the thought goes) in different micropatches different species will be more efficient. This helps explain how communities can retain diversity (as competitive superiority will not produce a
monoculture) and explains why more diverse communities are more
productive. They are more likely to include the species that are best
suited to the various local micropatches spread through the habitat.

Tilman's crucial theoretical idea is that of compensation. If one
population declines in numbers, another population, using somewhat
similar resources, expands, and hence stabilizes the overall productivity of the community. Importantly, the idea that populations compensate for one another's fluctuations does not depend on controversial
ecological assumptions. In particular, it does not depend on the idea
that population decline is caused by the competitive superiority of the
expanding species. We get community-level stability despite population-
level volatility because individual populations have somewhat overlapping resource requirements but quite different environmental tolerances. Tolerance differences explain why populations fluctuate out of
synchrony. When frost (for example) causes one population to contract,
the resources that the larger population used are now available (even if

only as space), and so another population can expand. The overall effect is to partially stabilize the overall productivity of the community. There is empirical evidence that supports this cluster of ideas. Tilman's own empirical work concentrates on Minnesota grassland plots, though he also reports African data supporting similar conclusions. Species-rich plots resisted drought better, overall biomass varied less in species-rich plots, and species-rich plots returned to the predrought biomass more rapidly than species-poor plots (Tilman 1996, 358; Tilman et al. 2006). In summary, Tilman argues that both theoretical models and experimental findings support diversity-stability hypotheses when these are taken to be hypotheses about communities rather than populations (Lehman and Tilman 2000; Tilman 1996, 1999; Tilman et al. 2005).

Thus the diversity-stability hypothesis has empirical support, and, most importantly, it is based on undemanding theoretical assumptions. There is a near-consensus in ecology that, in some measure, there is a positive relationship between diversity and stability (see the consensus report, Hooper et al. 2005). There is even some suggestion from the study of fossil reef systems that this diversity-stability relationship can be documented over very long time periods (Kiessling 2005; Naeem and Baker 2005).[11] However, there are problems. The experimental evidence in favor of the diversity-stability relationship depends on measuring plant biomass, but there are serious doubts about whether these ensemble relationships hold when we consider the interactions between plants and animals, and among animals. When our attention shifts to herbivores and those that eat them, resource exploitation efficiency may not be a stabilizing mechanism. To the contrary, enhanced resource use can cause overexploitation and hence productivity collapses (Loreau et al. 2001, 807).[12] David Wilson rightly points out that we should be cautious about inferences from individual efficiency to efficiency of the system as a whole. Evolutionary mechanisms reward actions that shrink the pie, so long as those that shrink it get a larger chunk of the smaller pie (Wilson 1997).

In short, though the diversity-stability hypothesis (and the related diversity-productivity hypothesis) is plausible, it is not yet demonstrable that more diverse communities are more stable (or productive). That is one reason to be wary of the conclusion that communities have causally important ensemble properties. There is a second reason for caution about this inference. Even if more diverse communities are more stable, it is not clear that they are more stable *because* they are more diverse. Diversity may be a symptom of causally relevant properties of individual populations rather than a causally important property of ensembles. We need to distinguish *redundancy effects* from *sampling effects*. First, redundancy. Suppose the overall productivity of the community depends on a set of key processes. These include the acquisition of energy by primary producers; the flow of minerals to and from the abiotic substrate; decomposition by the detrivores; and the flow of organic material from organism to organism via predation, herbivory, and similar activities. Ecosystem function depends on these key processes, and ecosystems that are more diverse, and hence have a variety of species with rather different tolerances that can compensate for one another driving these processes, are thereby more stable. If this is the right story, there is redundancy that matters in the system, and diversity itself is genuinely causally important (Naeem 1998).

However, there is an alternative possibility: the sampling effect. It may be that stability depends on the presence in the ensemble of some specific species. Suppose that what we know is that more diverse communities are more likely to resist exotic invasion. Resistance might depend on the presence of a species with a specific biological profile. All else equal, rich communities are more likely to contain such a species than a species-poor community. They have more tickets in the relevant biological lotteries (Wardle 1999). Variation in traits in the community is an ensemble property. If a community has an extensive range of phenotypes, many of which contribute causally to using the available resources efficiently (hence, for example, making it more difficult for potential invaders to establish), then the diversity of the community is itself causally crucial. Not so if diversity just increases the chance that a key taxon is present, and resistance to invasion depends on that taxon.[13]

So to establish an emergent property hypothesis, the covariation between the emergent property and its apparent effect must be robust, not limited to a few kinds of systems. And the relationship must be genuinely causal. Tilman's hypothesis and its relatives remain plausible, but no definitive case for the diversity-stability hypothesis has yet been made. The same is true of its close relative, the diversity-productivity hypothesis. Nonetheless, for reasons that will emerge in 6.6, we think the case for causally salient, emergent properties is very strong once our attention shifts to the effects organisms, populations, and functional groups have on the habitat in which they live. Some of these effects, it seems to us, are both critical to the biological profile of habitat patches, and are ensemble effects. Perhaps the clearest example is nutrient cycling: the processes through which crucial minerals flow from and to the soil; processes mediated by countless fungi, microbes, invertebrates, and plants. Moreover, these processes are synergistic rather than additive; in the recycling process the actions of one functional group create the input for another. To the dung, its beetle.

6.5 BOUNDARIES

If communities are ecological systems with causally salient properties, then, presumably, they have objective boundaries too. Thus if diversity really buffers a community against disturbance, there must be boundaries: a zone after which we stop counting, as addition of diversity *there* makes no difference to the extent of buffering *here*. Likewise, if communities are networks of interacting and self-regulating populations, there has to be a fact of the matter about which populations are part of a given network. But perhaps there are no such facts. Should we think of Karori Reserve as a single community? It is 252 hectares, and it is quite topologically varied. It is not large as the bird flies, and many birds that have been introduced to the reserve forage outside it. But for skinks, geckos, and most invertebrates, this is a sizable and diverse patch. So should we think of this as a single community, or as a multicommunity ensemble? The reserve's Web site implicitly treats it as two conjoined communities, one centered on a wetland, and the other on the regenerating forest. But skeptics doubt that these questions have objective answers (see, for example, Parker 2004).

At this point, it is important to set aside a potential confusion. Communities need not have sharp boundaries for them to have real boundaries. As with evolutionary biology, the existence of intermediate cases is no challenge by itself to the idea that—for example—communities are networks of populations whose demographic trajectories are under mutual influence. An extended family of white-winged choughs that live on the grounds of Australia's Commonwealth Scientific and Industrial Research Organisation (CSIRO), adjacent to, but foraging occasionally, on Black Mountain is somewhat influenced by events on Black Mountain. But by the interaction test, it may be neither a member of that community, nor not a member.

A more serious problem is the thought that populations will typically overlap rather than coincide, because the boundaries of particular populations will depend on their powers of dispersal, and these will vary from species to species. Roughly speaking, a population is a group of organisms of the same species that are potential mates (or rivals for mating opportunity). Mating capacity and mating rivalry depends on the mobility of organisms and their gametes. So consider Karori Reserve again. A local kaka population may overlap with a tui population, a local boobook owl population, and a number of skink and gecko populations. Moreover, if there are doubts about how to count communities in the Karori Reserve, those doubts will be much greater when we consider communities not bounded by sharp physical discontinuities. Black Mountain is much larger: thousands rather than hundreds of hectares. It is quite diverse. The topology is varied; fire has created habitat patchiness; there are important differences in microclimate; and some of it has been farmed until quite recently. But for the most part, there are no sharp changes as one moves across this patch, no abrupt differences that will matter to most of the species present, keeping local populations congruent with one another. There is not much reason to expect the dynamics of echidna populations to match those of larger and more mobile organisms, or those of smaller and less mobile ones. Black Mountain kangaroos may well compete directly for resources and breeding opportunities with kangaroos on the O'Connor Ridge (about a kilometer to the north). Echidnas are less mobile, so O'Connor Ridge echidnas are probably a source population for Black Mountain echidnas, buffering that group against population collapses rather than competing with them for scarce resources. Here is the challenge. Communities are systems with causally consequential problems only if they have objective boundaries. But they do not seem to have such boundaries.

Richard Levins and Richard Lewontin argue that community boundaries are defined by interaction patterns rather than sharp changes in physical conditions (Levins and Lewontin 1985, 138). On the Levins-Lewontin conception, communities are systems of strongly interacting populations, where "strong" and "weak" interaction are understood comparatively: the members of a community interact strongly with one another by comparison to influences on and from populations outside the community. In their view, communities are more or less closed networks of interacting populations. Their boundaries are zones in which interactions become fewer and weaker. They are zones in which biological events—local increases and dips in population—have less impact on the populations of community members. Even so, systems of strongly interacting populations will tend to occupy an identifiable physical space. Suppose populations of goannas, currawongs, and fairy wrens interact strongly, with the effects of goannas on fairy wrens mediated by their effects on currawongs (goannas are large monitor lizards that prey on currawongs. Currawongs are large corvids that prey on fairy wrens). In most cases, the strong interaction condition will imply that the territories of the three populations largely coincide. If they do not— if, say, the goannas and currawongs only intersect moderately—many currawongs will never encounter a goanna and vice versa. As a rule of thumb, interaction requires proximity. Think of such characteristic ecological interactions as predation, herbivory, mutualistic exchange of nutrients, and pollination. None of these are interactions at a distance.

Communities of strongly interacting populations will be roughly spatially identifiable.

Thus the view that communities are organized systems does not presuppose that they are bounded by zones at which biologically important abiotic conditions change markedly. Nor does it presuppose that we can always determine whether a given population is part of a community. But it does presuppose that patterns of interaction are clumped, that most populations are parts of networks whose members interact with one another more strongly than they interact with populations outside the network. It is far from obvious that this condition is typically met. It is quite likely that ecological interactions are not clumped in ways that enable us to identify bounded communities, even taking into account the fact that community boundaries are vague.

It is hard to tell just how serious this problem is, for there are ecological processes that can generate patchiness across a habitat. Organisms do not just passively experience their environment; they actively change it. Organisms in part construct their own niches (Odling-Smee et al. 2003). This is one mechanism (as Paul Griffiths has pointed out to us) through which an initially fairly homogenous territory can turn into a mosaic of quite different patches. Niche construction is one mechanism that can magnify an initial difference between patches, beginning a cascade that takes us from initially similar systems to a mosaic of quite different patches. Suppose, for example, that by chance eucalypts rather than acacias happen to initially predominate in one zone. Eucalypts have different environmental effects than acacias. They grow more slowly, but they live longer and eventually afford many animals homes in the hollows that form in them. They support very different pollinators; honeyeaters visit their flowers but not those of acacias. They produce very different litter. So an initial difference can generate quite marked differences between adjoining patches, thus generating two somewhat closed networks of interacting populations. We certainly cannot be confident that niche construction effects will increase landscape-scale heterogeneity, creating mosaic effects that bring populations of many species into rough spatial alignment with one another. But it is one possibility.

6.6 THE SPACE OF POPULATION ASSEMBLAGES

There is no definitive case for causally salient community properties. Regardless, we think that many ensembles have such properties. Organisms are profound agents of transformation both of their own and others' environments,[14] and recognizing that fact greatly strengthens

the case for thinking that communities are structured and have ensemble properties. Populations within a community can be linked via
niche construction networks. One population can influence another by
changing important features of the physical environment. Trees buffer
the wind and modulate the impact of storms while providing shelter
to many organisms (Jones et al. 1997). These indirect ecological links
expand the range of potential interactions in communities. Populations
act on one another via the physical changes they induce. Litter recycling
is the cleanest example. Plants produce litter as a by-product of their
life: fallen leaves, twigs, bark. A host of organisms live by consuming
the litter, and as a consequence of these actions, they return crucial
materials to the soil. This is absorbed by the vegetation, which in turn
produces more litter (Odling-Smee et al. 2003, 318–22). Moreover,
niche construction often involves ensemble effects. Soils are made organically, but not by any single population. A vast suite of very different animals, plants, and fungi make soils. Ants and other burrowing
animals turn over and redistribute soils; trees and other plants stabilize
soils; fungi, microbes, and a vast army of small invertebrates make soil
by consuming litter. Thus the discussion in 6.3 of community regulation
and of assembly rules understates the case for causally salient, system-
level properties by focusing so exclusively on density dependent forces,
of which competition and predation are the prime examples (Callaway
1997). The individualist view of ecology looks much more plausible
when niche construction is neglected.

We have contrasted the idea that communities are causally salient,
internally regulated, bounded systems with the individualist idea that
they are mere aggregates of overlapping populations that happen to
have fairly similar physical and biological tolerances (and hence are
merely phenomenological systems). But as the last few sections have
noted, we have here a spectrum of possibilities. The crucial factors that
distinguish an assemblage of indifference from a functionally organized
community come in degrees. Thus an assemblage may be internally
regulated to some extent. The power of internal regulation will depend
on the proportion of the component populations that interact strongly enough to influence abundance in the community, the strength of
those interactions, and their stability in the face of outside disturbances.
Likewise, an assemblage may have causally salient emergent properties
to some extent. Suppose, for example, that more diverse communities
really are more stable. But stability comes in degrees and in different
forms. The importance of the stabilizing effect of diversity will depend
on the degree to which diversity buffers the community against disturbance, the range of properties that are buffered against disturbance, and

the kind of disturbances whose effects are muted. Boundedness, too, comes in degrees; a network of interacting populations can be more or less closed; more or less spatially coincident rather than merely intersecting. Moreover, and most importantly, these factors may be partially independent of one another. If, for example, the stabilizing effects of diversity depend on compensation, an assemblage can have causally important emergent properties without being internally regulated.

We began this chapter by identifying phenomenological communities and noting that they play an important role as biodiversity surrogates. Identifying a community as a river wetland, an alpine grassland, or as a southeastern slopes eucalypt woodland gives us a reasonable guide to its alpha diversity. Identifying physically adjacent communities as phenomenologically distinct—a riverine community next to grassland habitat—likewise gives us a reasonable guide to their beta diversity. So in an undemanding sense, local ecological communities are units that we should recognize and count, and there are important differences between landscape-level ecological systems that contain many different phenomenological communities and those that are more homogenous. Beta diversity, for example, will be much higher, and invasion and perhaps other disturbances are less likely to spread uniformly through the landscape. Keeping track of phenomenological community diversity is likely to be predictively important for ecology and conservation biology, whether or not local communities are organized systems. We do not expect this claim to be controversial, so in this chapter we have concentrated on the much more controversial issue: whether local communities are units that must be recognized in causal explanations of ecological processes, and, if so, how they differ one from another. Hence our focus on the causal questions: are communities organized systems, with their own effects on their own membership and abundance? If so, do they differ systematically from one another in their organizational properties?

We are not in a position to answer this question, but we think we have developed a useful framework for its investigation. The features that make communities explanatorily salient come in degrees and are potentially independent of one another. These dimensions define an ecospace; different local communities will differ from one another in this space, not because of their phenomenological differences—differences in membership and physical environment—but because of their differences in causal organization. Thus we think it is productive to think of specific local communities as occupying differing positions in a 3-D space: the three dimensions being boundedness, internal regulation, and emergent property effects, rather than physical gradients like temperature or rainfall. These dimensions define a space of possibilities: specific ensembles at specific times and places will be more or less bounded; more or less internally regulated; have or lack important system level properties—properties like buffering against disturbance, nutrient cycling, fire-resistance—to some degree (as in fig. 6.2). A maximally indifferent assemblage in one corner of the space would consist of a set of populations that merely overlap, which do not significantly influence one another's demographic prospects, and which have no important collective impact on their environment. In the opposite corner of the space, there would be assemblages consisting of spatially coincident populations strongly influencing one another's demographic fates, and with important ensemble effects.

We like this way of representing the nature of communities, for it suggests an important research agenda. One set of questions will be about the circumstances under which communities of indifference become organized (and vice versa). Are sharp abiotic gradients important in aligning populations of different species or could niche construction effects also generate the right kinds of environmental patchiness? Disturbance, too, might play a role in generating patchiness; fire and flood are sources of sharp abiotic gradients. However, there is also a line of thought suggesting that disturbance can turn integrated communities into communities of indifference. There is suggestive (though far from decisive) evidence that Pleistocene communities are closer to being communities of indifference than earlier paleoecologies. Ecological as-

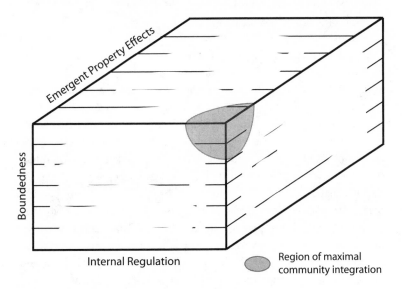

FIGURE 6.2. The space of population assemblages.

sociations between species were unstable over the Pleistocene, but they may have been less so in earlier environments (Valentine and Jablonski 1993; Coope 1994). The idea here is that the intensity of Pleistocene climatic fluctuations exceeded a threshold, causing community organization to break down, replacing regulated communities with unregulated ones. If true, this is clearly very important.

A second set of questions concerns the dimensions of ecospace. It's quite possible that our three dimensions already lump together organizational features that should be considered separately. For example, internal regulation might be achieved via niche construction, but it may also be the result of strong competitive interactions or obligatory mutualisms. It is not obvious that these characteristics should be aggregated into a single scale. Likewise, there may be a number of important and conceptually independent ensemble properties. The niche construction literature indicates that ensemble effects on physical features of the habitat may be important. For example, different suites of vegetation have markedly different effects on the water table and salinity of local systems. Western Australia, in particular, is suffering severe salinity problems because woodland was replaced by pasture, causing the water table to rise. Surface water then evaporates in hot dry seasons, leaving a salt residue. It may be that there are several dimensions here, not just one. Moreover, there are candidate dimensions we have not considered at all. One is openness to migration from the regional species pool. Island biogeography (and its more nuanced descendants) suggests that openness contributes importantly to richness and to stability (Ricklefs and Schluter 1993; Ricklefs 2004).

A third set of questions concerns the distribution of actual assemblages in ecospace, and in particular, whether there are correlations between phenomenological community types and location in our causal ecospace. Desert communities are phenomenologically similar, but are they in roughly the same space? Are all open woodlands? These questions offer an alternative approach to the vexed problem of the contingency of ecology we noted in 6.1. Even if the specific composition of communities is sensitive to the accidents of arrival and establishment, their structural properties may be predictable. Suppose, for example, that grasslands with different histories are nonetheless clumped in a particular volume of ecospace; imagine that they have ensemble properties but are not tightly regulated. This would indicate that in one important respect the processes that assemble grassland communities are not contingent. They build communities with similar structural properties, even if those communities have different members. Alternatively, the case for contingency would be strengthened if grassland communities were scattered through ecospace. One response to the difficulties involved in generating precise predictions about the changes in communities over time has been to argue that ecological theory was focused on the wrong spatial scale: ecologists should develop predictions about regional rather than local processes (Gaston and Blackburn 1999; 2000). A different response is to develop predictions about more abstract or global features of local communities rather than predicting the fates of particular actual and possible populations within them. In effect, we have seen this in Tilman's and Naeem's responses to May; they make predictions about overall productivity or stability rather than about the fate of particular populations. The suggestion here is a generalization of that approach to the contingency problem.

If individualism is essentially correct, information about species composition and numbers (together with information about the physical environment) captures ecological dynamics; there is no extra biological structure we need information about to explain ecological patterns. There would be no independent ecological ingredient to biodiversity. That fits the fact that ecological diversity has typically been characterized phenomenologically: by appeal to the physical parameters of the habitat (shallow, wave-influenced vs. deep water benthic communities) or by the predominant biota (pine forests, grasslands). As we have emphasized, such phenomenological characterizations are predictively useful. We can make a fair guess at what we can expect to see in a grassland. Grass, for example, will not be a surprise. But we think that characterizing local assemblages in terms of their positions in this abstract space opens up a research agenda in ecology in ways that phenomenological characterizations do not. It is a way of investing the open empirical possibility that there is extra biological structure that plays a central role in explaining ecological patterns. We think it is likely that there is such structure, and hence we need to go beyond species abundance and distribution information in explaining and predicting ecological patterns. But since we have focused entirely on one spatial scale in ecology, that of local ecological communities, we certainly have not proved that in this chapter. Indeed, we have not done much more than begin a preliminary sketch of this extra biological structure.

We now return to conservation biology, armed with an appreciation of the power of species-information-based measures of biodiversity, of the limits of that conception of biodiversity, and of the difficulty of supplemented species-based models in a disciplined and tractable way.

7 Conservation Biology: The Measurement Problem

In chapters 2–6, we focused on the explanatory and predictive significance of biodiversity properties, on their roles in driving important biological dynamics. We did not entirely neglect conservation issues, but we did not focus on biodiversity properties as targets of conservation policy. The example of the food web illustrates the important connection between biodiversity as cause and biodiversity as a policy target: biodiversity properties are targets of conservation policy because biodiversity properties, and changes in those properties, drive biological dynamics of fundamental importance. Identifying the causally salient features of systems identifies the sites in those systems at which interventions change outcomes (for an eloquent and detailed articulation of this view of causation, see Woodward 2003). Interventions can be deliberate human interventions, side effects of human activities, and (of course) disturbances that are entirely independent of us. Sunspots flare, volcanoes vent, faults shift, and soils erode independently of human action. So, for example, if individualist models of ecology of the kind we discussed in the last chapter are right, the policy implications are profound. On the one hand, individualism implies that ecological communities are predominantly modular, and hence the removal of one species is unlikely to have important consequences for most other populations. On the other hand, individualism also implies that these systems can be quite sensitive to perturbations in abiotic conditions; diverting water for irrigation, or allowing nutrient-filled runoff into a wetland might utterly transform it. So the causal and predictive considerations of the last few chapters are of great importance to conservation biology. These theoretical programs, when successful, identify levers of change in biological systems. But they cannot by themselves settle policy issues; they cannot tell us the human costs on intervention, and neither can they tell us what outcomes to aim for, and which to avoid.

In this chapter and the next, conservation biology becomes our central focus. In this chapter, we focus on measurement issues. These are difficult and controversial for two reasons. The first replays the theme of this whole book: measurement requires us to identify the explanatorily salient dimensions of diversity, because there will always be some way of comparing (say) one wetland to another that will count the first as the more diverse, and another procedure that will reverse the result. The point is the same as that made about the phenetics movement in systematics, and has the same rationale: there is no theory-neutral notion of overall richness any more than there is a theory-neutral notion of overall similarity. The second reason is that measurement procedures must be tractable. We must be able to measure features of biological systems even given the constraints on time, of resources, and information imposed on conservation projects. These resource limits seriously constrain measurement. As a consequence, conservation biologists almost never measure directly the full range of phenomena that they take to constitute the biodiversity of a system. Rather, they sample that diversity, or rely on measurable signs that vary (they believe) with biodiversity itself. Samples and signs are biodiversity surrogates, and this chapter will mostly be concerned with the evaluation of such surrogates.

While biodiversity and its protection is fundamental to the goals of conservation biology and the policies that discipline has devised, consensus on the importance of biodiversity has not been matched by consensus on the technical problem of biodiversity measurement. The last two decades have seen a proliferation of biodiversity measurement strategies, but a paucity of theory aimed at evaluating and comparing them. This proliferation is widely recognized in research volumes such as *Biodiversity: Measurement and Estimation* (Harper and Hawksworth 1995) and *Biodiversity: A Biology of Numbers and Difference* (Gaston 1996), and in textbooks on biodiversity such as *Biodiversity* (Lévêque and Mounolou 2003) and *Biodiversity: An Introduction* (Gaston and Spicer 2004). These works give thorough inventories of current measurement techniques, but are much less forthcoming on how measurement strategies ought to be compared with one another or how the success of biodiversity measurement strategies in general ought to be evaluated.

Formal conservation policy is even less useful than the technical literature in articulating a measurable concept of biodiversity. The United Nations Convention on Biological Diversity defines biodiversity in Article 2:

"Biological diversity" means the variability among living organisms from all sources including, inter alia, terrestrial, marine and other aquatic ecosystems and the ecological complexes of which they are part; this includes diversity within species, between species and of ecosystems.

These pieties treat "biodiversity" as a synonym for "all living things." Such a definition is of little use to conservation biologists trying to develop and evaluate methodologies for biodiversity measurement, and is of equally little use to conservation planning. Planning always involves choices, sacrificing one system to save another. So we begin this chapter by setting out a group of biodiversity measurement strategies. This is not a complete survey. We want instead to focus on how widely these strategies differ and on the considerations that are supposed to favor one rather than another. In the next chapter we move to a different set of issues: those involving costs and goals.

We begin our investigation of the place of biodiversity in conservation biology with a description of its use in current science, identifying the phenomena scientists actually measure when making judgments about diversity, and the phenomena they would measure if unconstrained by considerations of cost and effort. Once we turn to actual practice, we confront the problem of biodiversity surrogates noted above. We do not measure temperature by directly measuring the kinetic energy of particles and taking a mean. Instead, we use a substance (namely, mercury) with characteristics that are both highly sensitive to changes in temperature and that are easily measured. Analogously, it would be ideal to discover a sort of biodiversity thermometer. The strategy of using surrogates to detect biodiversity is the strategy of devising such biological thermometers, of identifying properties of biological systems that are reliable indicators of biodiversity properties. This strategy is almost universal in conservation biology, and many surrogates have been proposed. If conservation biologists are getting it right, these surrogates are reliable indicators of important characteristics of biological systems. Whether or not they are getting it right, these surrogates are reliable indicators of what conservation biologists take to be important about biological systems. So we now turn to a quick sketch of the most important surrogacy suggestions. As we shall see, there is a good deal of ambiguity about the status of these measured variables. Sometimes they are interpreted as signs of biodiversity, but not themselves as actual components of biodiversity. Counting family-level diversity in a system as a proxy for its morphological diversity exemplifies this approach. Sometimes they seem to be taken as representative samples, parts of the whole that indicate the whole. The use of indicator taxa exemplifies this approach. Counting butterfly species in two forests gives a component of species richness in each forest, and also can be used as a sign of the overall species richness of the two areas. Sometimes the measured variables seem to be taken to be a measurement of biodiversity itself, as in some views of genetic diversity.

7.2 COUNTING TAXA

We begin with the simplest idea, one that has been central to chapters 2–6. Perhaps we should measure biodiversity just by counting taxa, for the most widely used strategy for the measurement of biodiversity is counting taxonomic groups and estimating their frequency. These strategies typically distinguish between estimating alpha and beta diversity. The alpha diversity of a particular habitat patch is its local taxon richness (usually species richness): the number of taxa found in the community, weighted by abundance. A system with one very numerous species and a few rare ones is less alpha diverse than one in which the species are equally abundant (see Box 7.1 for details). The beta diversity is a relational measure; it measures the additional richness this patch adds to the regional system, and the species added to the count through surveying this community. Beta diversity (and its relatives) is very important to conservation planning, because that planning typically involves the selection of an ensemble of sites to maximize the overall protection of biodiversity. The difference between one community and others already protected (or considered for protection) is often as important as the intrinsic richness of a community.

As we have just noted, information about species richness is often joined with information about abundance; measures that combine information in this way include the Shannon Wiener Diversity Index and Simpson's Index (see Box 7.1). The intuitive background to such measures is the thought that a sample of (say) 100 organisms representing 10 species is not very diverse if 85 of the organisms belonged to a single species. If this were a plant community (for example), the characteristics of the community would depend largely on the phenotype of the hyperabundant species. Of course, the idea that ecological processes are controlled by the phenotype of the numerically abundant species can be trumped by special features of the rare species: if the hyperabundant species is an annual wildflower, and the other nine species are all species of large tree, we might well make no such assumption. Phenotype matters, and we will soon consider ways of making its importance explicit.

BOX 7.1: Diversity Indices

Diversity indices supplement species richness. The number of species represented in a sample (s) is supplemented with information about the evenness with which individuals are distributed between the species present. Evenness information is often represented as p_i (the fraction of individuals belonging to the ith species). Two common measures are:

Simpson's Index

$$D = \sum_{i=1}^{s} p_i^2$$

This is a measure of the probability that any two individuals in a sample will belong to the same species.

The Shannon Wiener Diversity Index

$$H' = -\sum_{i=1}^{s} p_i \ln p_i$$

This is a measure of the disorder of the sample (strictly the "entropy" as understood in mathematical information theory). On this measure, a highly diverse group is one with a great number of different types of individuals and roughly the same number of individuals of each type.

Counting species involves surveying (perhaps several times to account for seasonal variations) the organisms in a particular habitat, and sorting the specimens collected into species. One advantage of this strategy is that, for some taxa it is relatively simple. Because organisms of different species tend to be morphologically distinct, workers with limited training in taxonomy can roughly estimate the number of species in an area. Estimates of species numbers made by those without formal taxonomic training will be "rough" because they will be confounded by cryptic species (populations that do not interbreed despite a high degree of morphological similarity), radical sexual dimorphism (species in which males and females are so different as to appear to be members of different species), and radical morphological differences in successive life stages (common among invertebrates). Moreover, our ability to distinguish between species is much more reliable for some taxa (for example, vascular plants and vertebrates) than others (for example, fungi and protists) (Berlin 1992). So while there are practical advantages to species counting, there are practical disadvantages as well.

The vertebrates and vascular plants in a region can usually be identified fairly accurately, but the same is not true of invertebrates, fungi, and microbes, and these are important components of taxonomic richness. Abundance is difficult to estimate reliably, too. Hence conservation biologists often use proxy taxa, like bird diversity, as indicators of overall taxonomic diversity, and of changes in diversity.

Counting species is also theoretically well motivated. As we have argued in chapters 2–6, if there is a decent candidate for a good overall measure of biodiversity, a measure relevant to many of the theoretical and practical projects of the life sciences, it is based on the species richness of a biota. Despite the controversy over species definitions, there is widespread agreement that species are objective features of the biological world: species are the crucial units of evolution. Moreover, as we have noted already, there are natural ways of supplementing information about species richness. We can add abundance data. In chapters 2–6, in talking about species richness as an overall measure of biodiversity, we talked of information about the species and their genealogies. So we can add phylogenetic information, to represent the difference between a biota that represents a number of ancient clades, and a biota dominated by a large population of recently evolved close relatives. The small mammal fauna of Tasmania contrasts with that of North Queensland in this regard: both are diverse, but North Queensland has a large number of recently evolved true rodents, where Tasmania has more representatives of ancient marsupial lineages.

However, while in principle it is possible to supplement a species-richness-based account of biodiversity with phylogenetic information, in practice it is not obvious how to do this in a precise and tractable way. This problem is particularly pronounced in estimates of beta diversity. While we might plausibly estimate the total species count of a large region, it would be much more difficult to estimate a phylogenetically adjusted account of its species diversity. As we have remarked, almost all biologists share the judgment that different species represent different amounts of biodiversity. The two surviving species of tuatara (genus *Sphenodon*) are remarkable both morphologically (for the possession of a hidden third eye) and phylogenetically (as the last survivors of the order Rhynchocephalia (Sphenodontia), sister group to the snakes and lizards). Given this, many think that conserving a species of tuatara represents a much greater saving of biodiversity than, say, preserving a species of minnow.

The tuatara are such classic examples of "living fossils" that they make the intuition that species are not all equally unique very vivid. But we do not need such a vivid example to make the point, as is shown by a thought experiment of Harper and Hawskworth (1995, 7). They suggest that we consider how much biodiversity is present in a series of hypothetical sites. Each site contains just two species. One is a species of *Ranunculus*, a genus of flowering plant within the buttercup family (Ranunculaceae), and the other is:

1. Another species of *Ranunculus* from the same section of the genus.
2. Another species of *Ranunculus* from a different section of the genus.
3. A species from a different genus in the same family (Ranunculaceae).
4. A species from a different family within the same order as the Ranunculaceae.
5. A species from a different family and in a different order (for example, a grass).
6. A rabbit.
7. A fungus of the genus *Agaricus*.
8. A protozoan of the genus *Amoeba*.
9. An archaebacterium.
10. A eubacterium of the genus *Pseudomonas*.

In some important sense of biodiversity (the thought goes), these samples are not equally biodiverse. As Robert May puts it:

> One of the basic conceptual issues in quantifying biological diversity is the extent to which a "species" does or does not represent the same unit of evolutionary currency for a bacterium, a protozoan, a mite, and a bird. (May 1995, 15)

Thought experiments like these have led ecologists to search for a measurement strategy that more accurately reflects the differences among organisms. We need some representation of species structure, not just the numbers of species present. Family-level diversity is sometimes suggested as a surrogate for this structure. So some taxon-counting measures of biodiversity count families instead of, or as well as, species. The family is a common choice because families are less subject to taxonomic revision than genera and they are more informative than more inclusive taxonomic levels such as orders and classes. This is one way we can, in practice, add information about the evolutionary history and morphological disparity to our measure of biodiversity. A biota that includes ten families of arthropod represents more evolutionary history and disparity than a biota that includes two. That said, we have already

seen the serious limits on the use of higher levels of the Linnaean system to capture biodiversity. There is no robust scientific theory that
allows us to settle disputes about whether a particular group of taxa
constitutes a family or not. This is not to say that we could pick any
assemblage of species and call it a family (at the very least such groupings must be monophyletic). As a clade grows by speciation at the tips,
the tree of species so formed gets larger and larger. Within any large
tree, there will be many branches that we could pick out and name, but
that science has chosen not to name. Perhaps in a rough-and-ready way,
family-level diversity is a surrogate for phylogenetic diversity. But this
will be at best a rough measure. Conservation biologists influenced by
cladism have tried to do better.

7.3 MEASURING PHYLOGENETIC DIVERSITY

The great theoretical strength of cladistics is that it does latch onto
something real in the world: phylogenetic structure, the massively
complex set of relationships that is the "genealogy" of species. There
is nothing conventional or subjective about the claim that a bat and a
bear really are closer phylogenetic relatives than are a bat and a bee.
It's not surprising then, that many have sought to exploit this fact
about nature in the measurement of biological diversity. Instead of
relying on intuitive judgments of phenotype distinctiveness, one of the
aims of those who want to measure biodiversity directly from cladistic
principles has been to try to devise a measurement strategy that treats
all speciation events as contributing equally to biodiversity. There is a
wide range of strategies available, but the most widely used[1] measure
of phylogenetic diversity is due to Daniel Faith (1994). However, see
also Owens and Bennett (2000), Posadas et al. (2004), and Barker
(2002).

One aim of the strategy is to pick out the group of species (from a
larger group being studied) whose members are most distantly related
to one another. To do so, Faith defines closeness of phylogenetic relationship in terms of the number of speciation events that separate a
group of taxa. So, for example, two sister species are separated by one
event. The direct offspring of those two sister species are separated by
three, and so forth. So we might think of the basic strategy as tracing a
line between taxa on the phylogenetic tree and counting the number of
nodes (that is, speciation events) along that line. The other aim of this
strategy is to try to capture the rate at which particular lineages evolve.
One of the reasons why phylogeny is not a perfect predictor of phenotype is that species evolve at different rates. So if two sister species

experience very different selection pressures, then one may evolve much faster than the other and thus end up looking much less like the parent species than its sibling. Faith thinks that we ought to take this evolution between speciation events into account when measuring biodiversity. He proposes to plot the evolution of character states onto the phylogenetic tree. When we trace the line between taxa, we can count not just how many nodes we pass but also the number of character states that have evolved along the way.

To calculate Faith's phylogenetic diversity we must first construct a cladogram that includes feature information (information about character state changes that occur either at or between speciation events). An example of such a cladogram is given in figure 7.1. The idea behind phylogenetic diversity is that if, for example, we could save some but not all of the taxa shown in figure 7.1 then we would set about this task by looking for a "minimum spanning path." Assume that we only have funding sufficient to save four out of the ten taxa shown. We then find all the paths on the cladogram that connect four species and choose the path out of that group that includes the greatest number of speciation events as well as the greatest number of character state changes. Those four species are the ones we should save.

If this seems a bit abstract, analogy might help. Think of figure 7.1 as a road map. At the tip of each branch is a destination and each of the dots represent potholes. The minimum spanning path is just the bumpiest way of getting to a given number of destinations. The minimum spanning path for the tree in figure 7.1 is shown in figure 7.2. Despite acknowledging phenotypic difference, this is explicitly a cladistic theory. Faith argues (1994, 4) that the advantage of using phylogenetic diversity

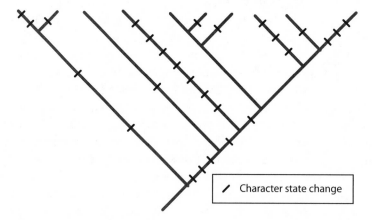

/ Character state change

FIGURE 7.1. Cladogram with character state changes. This diagram depicts the ancestry of ten extant species.

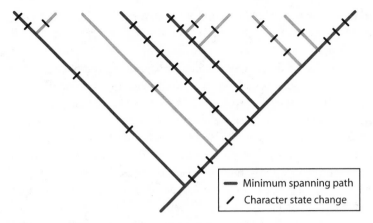

FIGURE 7.2. The minimum spanning path.

based on minimum spanning paths is that it will count traits that are structurally identical, but that result from evolutionary convergence, as different traits.

However, whatever the in-principle merits of Faith's proposal, in all but the simplest of cases it does not seem practicable. Most existing cladistic analyses do not contain the amount of information required for a measurement of phylogenetic diversity that includes comprehensive information about phenotypic difference. Moreover, as we noted in our early discussion of phenetics, the notion of *complete* information about phenotypic difference, is itself ill defined. So there will be difficult choices to make in deciding which information to include. Furthermore, cladistic systematics is increasingly dominated by cladograms derived from molecular data. Of course we can treat molecular change as character state change, but given that molecular difference does not covary cleanly with phenotypic difference, we cannot base our measure of phylogenetic diversity on both types of data. Faith's method might still be an ideal toward which we might work, but it would be vastly more labor intensive than species or family counting.

Moreover, and most importantly, despite the in-principle objectivity of the method, it is theoretically unmotivated. What exactly would distinguish a regional biota that was more Faith-diverse than one that was less Faith-diverse? Would it show more evolutionary flexibility on short or long time scales? Would it provide more resilient ecosystem services? Would it be more phenotypically disparate? If Faith-diversity is a measure of a causally important dimension of biological systems, we need an explicit case for that view. Equally, if Faith-diversity is a goal, a measure of some valuable feature of biological systems, that case must be explicit too (we will see a sketch of such a case for a measure similar to that of Faith in the next chapter). Measurement strategies need to be explicitly linked to claims about value or claims about intervention points.

Phylogenetic diversity is a blend of phenotype and phylogeny, but it is not a satisfactory blend. It is committed to the view that every character state change is of equal importance in measuring biodiversity, and that is no more plausible than the idea that every species is of equal importance. Given the fundamental implausibility of this view one might expect to see purely phenotypic measures of biodiversity, and indeed such approaches have been advocated (for example, Roy and Foote 1997). However, they are not common methodological choices for reasons that we discussed in chapters 3 and 4; once we abstract phenotype differences from a phylogenetic context, we have lost the most objective way to choose the traits to measure and compare. So we shall suggest that one option worth considering is the use of local morphospaces to explore the fate of a clade in different regions. We could, for example, compare the phenotypic diversity of New World versus Old World monkeys or Australasian versus American parrots using such local morphospaces. The common history of the clade makes them phenotypically commensurable; we can use the same dimensions to plot their spread in a common morphospace. Theoretical morphology is an important tool for thinking about biodiversity differences, but only in combination with genealogical information about the history and relationships of species.

7.4 MEASURING GENETIC DIVERSITY

Genetic diversity is crucial to conservation biology. As we noted in 5.1, populations on the brink of extinction often have too little genetic diversity. Selection pressures that would simply delete unfortunate phenotypes from larger populations may well destroy small populations because they lack the variations that would allow them to respond successfully. Moreover, measuring genetic diversity certainly has methodological attractions. DNA sequences are relatively easily identified, and the differences between sequences are more discrete and therefore more countable than phenotypic characters. A new and important research effort aims at identifying DNA bar codes, short DNA sequences that show little within-species variation compared to their variation between species. There has been some success in identifying a characteristic class of such sequences of animals; the situation with other taxa seems less promising. If we can find such bar codes, they will be an important tool for taxonomy and hence conservation biology, revealing the presence of sibling species, and enabling field workers to identify morphologically cryptic organisms. Many invertebrates have life cycles that involve stages that do not advertise their specific identity (Savolainen et al. 2005). Perhaps the most promising role for studies of genetic diversity is in understanding microbial diversity. Importantly, we can sample and amplify the DNA in a substrate, and thus get some information about both the variety and number of microorganisms present in the environment from which the substrate has been extracted. This technique has been used to estimate microbial diversity and community organization in environments as different as soils, human guts, and the open ocean (Falkowski and de Vargas 2004; Fierer and Jackson 2006; Gill et al. 2006). There are many uncertainties about these methods because the fragments of DNA that are amplified have to be assembled into putative organism genomes. Even so, measuring genetic diversity is a window onto an important aspect of biodiversity that is largely invisible to other methods for its assessment. These uses of DNA bar codes are uncontroversial. Much more controversial is the idea that DNA bar coding can largely replace traditional systematics. We agree that this more ambitious aim for DNA bar coding is wrongheaded; DNA bar codes need to be calibrated against an independently identified species phylogeny (Herbert and Gregory 2005; Smith 2005; Will et al. 2005). As always with a biodiversity surrogate, we can never just assume that there is a reliable relationship between the indicator property and the target property.

So there are good reasons to focus on measuring genetic diversity within biological systems. Genetic diversity is causally important (it is certainly part of the real diversity of biological systems) and it may covary well with other important aspects of diversity. Genetic similarity is certainly a reasonable predictor of important phenotypic similarity (Williams and Humphries 1996, 57). But there are also confused reasons; in particular, the idea that genetic diversity is fundamental and other dimensions of diversity are not. This confuses a surrogate for biodiversity with diversity itself.[2] For example, James Mallet argues:

> Biodiversity consists of the variety of morphology, behaviour, physiology, and biochemistry in living things. Underlying this phenotypic diversity is a diversity of genetic blueprints, nucleic acids that specify phenotypes and direct their development. (1996, 13)

It is certainly true, as we have noted, that the biochemical structure of genetic material provides us with quantifiable differences. But base pair similarity and difference is one thing; gene similarity and difference is another. Functioning genes are typically in the range of hundreds to thousands of base pairs. Furthermore, some portions of our genomes appear to play no protein-coding role in the development of phenotype, though it is increasingly likely that much untranscribed DNA has a regulatory function. Given this, it is at least theoretically possible for two species to display a high degree of similarity with respect to base pairs without sharing many genes. Moreover, the relationship between genotype and phenotype is complex. We discussed some of those complexities in 5.2–5.4; another symptom of that complexity is the so-called C-value paradox, the fact that there is so little relationship between genome size and (apparent, intuitive) morphological complexity. The variation in genome size, and its lack of connection with phenotype complexity is really quite striking. Genome size varies by a factor of 200,000 in eukaryotes (Ryan Gregory 2001), and not because some eukaryotes are small and simple and others are huge and complex, as the following data (taken from Zimmer 2007) show:

GENOMES SIZE FROM SMALL TO LARGE

Nematode (*Caenorhabditis elegans*): 100 million bp (bp = base pairs)
Thale cress (*Arabidopsis thaliana*): 160 million bp
Fruit fly (*Drosophila melanogaster*): 180 million bp
Puffer fish (*Takifugu rubripes*): 400 million bp
Rice (*Oryza sativa*): 490 million bp
Human (*Homo sapiens*): 3.5 billion bp
Leopard frog (*Rana pipiens*): 6.5 billion bp
Onion (*Allium cepa*): 16.4 billion bp
Mountain grasshopper (*Podisma pedestris*): 16.5 billion bp
Tiger salamander (*Ambystoma tigrinum*): 31 billion bp
Easter lily (*Lilium longiflorum*): 34 billion bp
Marbled lungfish (*Protopterus aethiopicus*): 130 billion bp

Indeed, Ryan Gregory points out that the 200,000-fold range is found across single-celled eukaryote lineages; the genome of *Amoeba dubia* is more than 200,000 times larger than that of the microsporidium *Encephalitozoon cuniculi* (Ryan Gregory 2001, 66).

In the light of this complex relationship between genome and phenotype, it has increasingly been argued that it is misleading to think of the genome as a program that controls or organizes development (see, for example, Gerhart and Kirschner 1997; Oyama et al. 2001). While the genome does direct development, it doesn't do so alone. A host of behavioral, embryological, and environmental resources are required for the development of an individual, and changes in these factors can produce radical differences in the developed individual (for a comprehensive

survey of these phenomena, see Jablonka and Lamb 2005). The maintenance of stable and diverse global gene pools is an invaluable tool in the
fight to achieve stable and diverse global ecosystems. Moreover, measuring genetic diversity gives us some insight into the otherwise hidden
world of microbial diversity and community structure. Finally, there are
genuine measurement advantages in focusing on gene diversity; it is an
important diversity surrogate. That said, we see no reason in general to
equate biodiversity in conservation biology with genetic diversity.

7.5 BIODIVERSITY SURROGATES

Biodiversity surrogates, in all probability, do not vary independently
from one another. There is clearly an important correlation between,
for example, species richness and family richness.[3] Nonetheless, the
various measurement strategies rest on different foundations. Some tie
biodiversity to speciation. Others tie it more closely to phylogenetic
structure. Some include a morphological component. Others come
close to tracking common intuitions about biological diversity. But
measurement strategies in conservation biology have to be especially
responsive to tractability issues; often conservation biologists measure
what they can, with the expectation (or hope) that the facts that can
be measured in the field track those believed to be of causal importance. It has long been recognized that conservation biology is a "crisis
discipline" (Soulé 1985). Its raison d'être is to be found in overpopula-
tion, intensive exploitation of environmental resources, habitat loss,
and pollution. These factors lead to species loss and environmental
degradation. Global conservation is a daunting task performed by too
few people and with insufficient funds. These facts constrain methodology. Conservation biologists must therefore concentrate their efforts on
"what is feasible, what is too crude to be useful, and what is unnecessarily detailed" (Fjeldså 2000).

Resource constraints sometimes bite very hard indeed, and hence
there are simpler and cruder surrogates than species richness. As
organisms tend to be specialized to niches in which they occur, as a
rough regularity (since it ignores generalists), different niches will likely
be filled by different organisms. The greater the difference in niche, the
more the occupants will differ in their genetics, morphology, and behavior. As we noted in the last chapter in discussing the value of phenomenological communities as a guide to beta diversity, we can use features
of environments as surrogates for the biodiversity that inhabits those
environments. So, for example, environmental parameter diversity
rests on the assumption that any available niche will be occupied by

at least one species (for a good discussion of this rather complex idea, see Sarkar 2002, 142–43). What it measures is diversity with respect to niches, but (if the basic assumptions are correct) what it detects is biodiversity.

There are even cruder measures: using satellite photography to estimate vegetation cover, and treating this as an index of biodiversity and biodiversity change. These measures are crude, but one of the main worries of those concerned with the conservation of biodiversity is the impracticality of strategies that involve the measurement of large numbers of properties of vast numbers of organisms. That is why we returned again and again in chapters 2–6 to the idea of phylogenetically enriched species information as a surrogate for biodiversity in general. It is a plausible compromise between what we would like and what we can do. Typically, here is information about species present in biological systems, and traditional taxonomy still encodes a lot of information about the genealogy of a species for all its subjectivity, failures to include stem species, and its use of paraphyletic groups (dinosaur, reptile). Thus a good flora and fauna (supplemented by some rough-and-ready abundance data) provides a sensible starting place in any study of biodiversity (where we are otherwise uncommitted to the nature of the diversity that is driving the system in question).

It is one thing to estimate the diversity of a system; it is another to be confident that the system continues to be as diverse. Even using surrogates, estimating diversity is often difficult and expensive, and yet systems are in a state of flux, and we can rarely assume that they are in equilibrium. Conservation biology badly needs surrogates for detecting change in previous baseline states. It is common to use proxy taxa to detect change. The idea here is to detect disturbance and estimate its severity by using change in abundance of some indicator taxon: a canary species whose loss or decline is a good indicator of general loss or decline. Thus an ideal indicator taxon is one that is very sensitive to habitat change, can easily be surveyed, and whose taxonomy and natural history are well known. Invertebrates make particularly good indicators as their short life spans mean that a change in breeding rates is easy to detect (Greenslade and Greenslade 1984).[4] But, clearly, even if there are indicator taxa in a habitat, they are difficult to identify with confidence, for (as with all surrogacy methods) the use of indicator taxa involves extrapolation from observed facts about the ecologies of known taxa in studied environments, to predictions about biodiversity in different environments under different conditions.

In 7.1 we noted that a good surrogate must be both practically usable in the field and a reliable indicator of its target property. In his recent introduction to the philosophy of conservation biology, Sahotra Sarkar discusses surrogacy extensively as part of his defense of the idea of ranking places according to their *relative* biodiversity value (Sarkar 2005, chap. 6). Sarkar thinks it is neither necessary nor possible to give an explicit definition of *absolute* biodiversity. Instead, he suggests that biodiversity can be implicitly defined by a ranking procedure using surrogates, a procedure that takes into account both the objective biological richness of places we have identified as candidates for protection and the practical constraints on our abilities to measure and protect this richness. Sarkar accepts the idea that there is an element of choice in the selection of surrogates, but we think he understates the problem of evaluating surrogates. In our view, we can assess the adequacy of surrogates only by explicitly addressing the question: what aspects of biological richness do we wish to conserve, and why? Butterflies, for example, have prima facie advantages as surrogates because (as with birds) natural history enthusiasts have generated a good database about their abundance and distribution. Moreover, as adults, they are readily identifiable. Butterfly richness may be a true surrogate for species-level taxonomic richness. But by itself, that does not tell us that butterflies are a good surrogate for other aspects of biodiversity. It is true that conservation biology would not have to address this problem if we knew that the various kinds of biodiversity covaried well with one another, if phenotypic distinctiveness covaried well with ecological complexity, which covaried well with levels of endemism or with phylogenetic distinctiveness. If various versions of biodiversity covaried, a good surrogate for any form of diversity—for example, species richness—would be a decent surrogate for all the others. But we do not know that (for some initial reservations in the conservation context, see Andelman and Fagan 2000).

The role of surrogates and index species has added both complexity and confusion to the literature on biodiversity in conservation biology. As we complained in 7.1, it is often not clear whether the features of a system being measured are seen as direct measures of target properties or whether they are surrogates: measurable proxies for causally relevant properties. In some cases the situation is unambiguous. The recent and increasing use of satellite images to assess the extent of vegetation cover in making conservation assessments is clearly the use of a mere surrogate. This technique is chosen because the data are easily available, not because anyone thinks we are thereby directly measuring the biodiversity that matters (see, for example, Margules and Pressey 2000). In contrast, Faith's phylogenetic diversity is probably conceived as a measure of the target property itself. But in other cases, the profusion of surrogates has led to much confusion, as our discussion of genetic diversity illustrates. Mallet tells us that "the diversity of life is fundamentally genetic" (1996, 13), whereas Williams and Humphries (1996) talk as if genetic diversity is better thought of as a surrogate for biodiversity, particularly in conservation settings.

Further, there is often little calibrating information about proxies and their reliability. This is no accident. They are used because it is difficult to get direct information about the causally relevant target properties of the system. That very fact makes proxies difficult to calibrate. For example, the coevolutionary interactions between butterflies and flowering plants probably make it safe to assume that areas rich in butterfly species are species rich. But it is not safe to make the converse assumption: that butterfly poor patches are species poor. So there are severe practical problems in calibration. But conservation biology faces theoretical problems in choosing target properties: we cannot choose what properties to conserve without an account of conservation aims. The literature is often not explicit (as we saw in discussing Faith diversity) on why particular target properties are chosen. To make further progress on this issue, we finally have to move beyond purely empirical issues about the driving properties of systems to claims about goals of conservation biology.

8 Conservation Biology: The Evaluation Problem

In chapter 1 we argued that the concept of biodiversity has to be made precise by tying it to specific scientific enterprises. The fact that, for example, species richness is commonly used as a measure of biodiversity in conservation biology does not imply that the maximization of species richness is an appropriate goal for conservation biology. That could only be established by a further argument demonstrating the scientific relevance of species richness and variation in species richness. We do not think that measurement strategies in conservation biology have been convincingly connected to wider theories that show the importance of the magnitudes measured.

In chapter 1 we outlined two broad reasons for interest in biodiversity magnitudes. We can track biodiversity as a signal of the processes that produce it. Alternatively, we can focus on the consequences of diversity. Conservation biologists are interested in the processes that generate biodiversity, but typically because they want to use information about those processes to intervene in biological systems. They want to conserve biodiversity. But why is that an important goal, and which aspects of biodiversity? This question leads us naturally to the problem of value, and to environmental ethics.

There is an important link between environmental ethics and conservation biology. Ideally, the former tells us what to conserve and the latter tells us how to conserve it. This book is about science, not ethics, and we shall address ethical issues only to the extent that they make a difference to scientific theory and methodology. In practice, this allows us to set aside a large portion of environmental ethics,[1] because much of this is irrelevant to our purposes. Let us explain.

8.2 IS BIODIVERSITY INTRINSICALLY VALUABLE?

If environmental ethics is to be relevant to conservation biology, it must address the value of ecosystems and their components, and do so in a way that is tractable and commensurable. As Sahotra Sarkar has emphasized, much of conservation biology involves assessments of relative importance. However, a group of theories in environmental ethics cannot be yoked to this task, and so they can be discounted from our investigation. One is the idea that ecosystems and their components are intrinsically valuable.

This idea enjoys wide support. The preamble to the United Nations Convention on Biological Diversity states that biodiversity is intrinsically valuable. This is an attractive idea to many people, as it reflects the sentiment that we care about nature not as resources[2] ripe for harvest, but rather as a good in itself; we are stewards responsible for taking good care of the world of life rather than owners free to dispose of it as we wish. This intuition is the basis of Aldo Leopold's (1949) claim that actions that harm the environment are wrong independent of the effects they might have on the interests of humanity. That idea in turn has become the central tenet of deep ecology. But for all this popularity, the idea that biological systems have intrinsic value poses important difficulties for those who seek to integrate environmental ethics with scientific practice.[3]

We normally think of value as linked to, and dependent on, evaluation. Something is desirable because agents do, or might, desire it. Something is valuable because agents value it. Theories of intrinsic value seem to cut this link. To say that biodiversity is intrinsically valuable is to say that it would be valuable even if nobody were to actually value it. Indeed, it would be valuable even if there were no sentient beings that could value it. This conception is typically defended by "last agent" or "no agent" intuitions; we are (for example) invited to share the intuition that a supernova that wiped out a world of rich, flourishing life would be a tragedy, even if no sentient agent had ever evolved at or moved to that world (Norton 2003, 164). Even if we find those intuitions persuasive, accepting their message need not completely cut the tie between value and evaluation. We can think the biodiversity of the lost world is valuable not because it is valued by actual agents, but because it would be valued by a rational agent were he or she to observe the nova unfolding and the blast of radiation sweeping brutally through the system. These "ideal observer" theories of value are currently quite popular (see, for example, Michael Smith's *The Moral Problem*). So our problem with intrinsic value theories is not with the idea of intrinsic value as such but with the tractability and commensurability of this conception of the value of biodiversity. Perhaps the intuitions generated by contemplating in the imagination these unexperienced disasters are robust enough to show that living systems have some value independent of agents' actual evaluations (indeed, we think this ourselves, as we shall show in 8.4). But they are surely not robust enough to establish comparative judgments, or to show which aspects of biodiversity are of special importance. Asking people to report their intuitions about events that would happen after their death as the last person in existence is rather like asking people's intuitions about what it would feel like to be made of cheese. The premise is too far removed from ordinary experience. Once we notice the many dimensions of biodiversity this epistemic problem becomes worse.

8.3 DEMAND VALUE

The most plausible strategy comes from broadly utilitarian theories of environmental ethics, that is, from theories that tie the moral worth of an action to its effects on the maximization or minimization of some natural property. Some versions tie value to the maximization of pleasure, happiness, or preference satisfaction. Others tie value to the avoidance of pain, unhappiness, or frustration. The simplest such theories equate the value of ecosystems and their components with the resources and services those things currently provide to human populations; they have a "demand" value that warrants the considerable investment required for their conservation. This family of theories has problems of their own. All versions of utilitarianism face the problem of aggregating individual cost benefit trade-offs into a collective assessment. This is difficult because benefits to some impose costs on others. Conserving the forest around a watershed to protect the delivery of clean water to those downstream will advance one set of interests, but at a cost to those who would have benefited from the resources that are locked into the forest. It is difficult because different individuals evaluate the same situation quite differently: taipans are charismatic megafauna to us, a terrifying menace to those phobic about snakes. That said, some of the benefits derived from biological systems both accrue to large numbers of people and are uncontroversially central to well-being. Many species are of obvious and undisputed importance. Some provide food or medicine or industrial resources. Some are of great ecological importance. Natural ecosystems provide many crucial ecosystem services: clean water; the protection of river systems from salination, erosion, and pollution; and they recycle nutrients and sequester carbon. They help stabilize weather and climate; they help make the free oxygen we breathe.[4] Others are just fortunate to be members of the "charismatic megafauna." These are medium to large organisms that humans find attractive or exciting, such as whales or tigers. These privileged organisms make regular appearances on the Web sites of organizations promoting conservation. They do, however, raise the aggregation problem: some of us rate the conservation of these animals as of high importance; others would give them little, or even negative, weight.

But many organisms do not fit any of the categories just mentioned, and this presents a problem for those who deplore the cherry-picking approach to nature conservation and advocate in its stead wholesale conservation of the natural world. As Elliot Sober puts it:

> The problem for environmentalism stems from the idea that species and ecosystems ought to be preserved for reasons additional to their known value as resources for human use. The feeling is that even when we cannot say what nutritional, medicinal or recreational benefit the preservation provides, there still is a value in preservation. It is the search for a rationale for this feeling that constitutes the main conceptual problem for environmentalism. (1986, 173–74)

As we have argued at length, there is more to diversity than species richness. But species bring out Sober's challenge well, for many species are not distinctive. They are very similar to many other closely related species with which they share many morphological and ecological characteristics. A good example is the snail darter whose plight we discussed in chapter 1, just another minnow that was neither economically, ecologically, nor aesthetically important. Indeed, an ultimate public relations handicap is faced by many species because they are yet to be discovered by science. But for those who have been judged and found unexciting, ought we be entitled (as Sober suggests) to engage in rational attrition? The snail darter problem is especially pressing because much of the demand value satisfied by biological systems consists of ecosystem services. Ecosystems protect water supplies, they stabilize and renew soils, they are sources of fuels and wild foods, they moderate the impact of storms, and they store carbon. While there is persuasive evidence (that we will shortly discuss) that species-rich systems deliver these services more reliably than species-poor ones, these services typically do not depend on the presence of specific species, especially not rare, narrowly distributed species. The species at most risk are those least likely to have a high demand value in virtue of their contribution to ecosystem services.[5]

In short, a demand value model of biodiversity conservation has important virtues, but it is also challenging. Demand value is scientifically corrigible in the right way. It enables us to assess the relative worth of different regions, and it would lead us to place a high value on protecting the basic ecological mechanisms on which we depend. There is some evidence that this should lead us to have a strong interest in conserving some rare species and not just the large and obvious components of important ecosystems (Lyons et al. 2005). But there is no reason to suppose that it would lead us to place a high value on every vulnerable species, or on many small and isolated ecological associations, however distinctive. This is because it does not tie value to diversity per se. Rather, it ties it to specific uses: importance as a resource, crucial ecological function, or to the rather more nebulous attribute of being much loved by the general public. Perhaps this is just the right answer, albeit not a very green one. For conservation biology, the biodiversity that matters is just those properties of biological systems that make them reliable providers of ecosystem services. Ecosystem services will include aesthetic and recreation services and hence the biota we value and use directly: the megafauna, coral reefs, and the like. If so, the Tennessee congressmen were right to kiss-off the snail darter. We are, however, not forced to this conclusion. We do not have to choose between theories that lack a strong epistemic foundation or a demand value that sees a great number of species as being of little value.

There are alternatives. For one thing, we do not have to take actual human values as fixed. Bryan Norton presents the idea of transformative value as a means of countering those whose demand values center more on consumables than on environmental amenity. Transformative value is roughly the value that we would see in nature if we had more rational preferences (where rationality is largely judged in terms of self-consistency; see also Sarkar 2005). In Norton's terms:

> This more complex, though still anthropocentric, value system is doubly congenial to the goals of environmental preservationists. It allows them to express their legitimate concern that runaway expansion of human demand values, especially overly materialistic and consumptive ones, constitutes much of the problem of species endangerment. It also highlights the value of wild species and undisturbed ecosystems as occasions for experiences that alter those very felt preferences. (1987, 511)

We agree with the basic premise that demand values might look quite different if they were the result of more rational reflection. Even so, unless we make some strong and controversial assumptions about rational reflection and the evaluations that such reflection generates,[6] this will not solve Sober's dilemma. Given our current state of scientific knowledge, even with considered rational reflection, many species simply appear to be surplus to requirements. Transformative value, like demand value, does not tie value to diversity, but to specific elements within it.

8.4 THE OPTION VALUE OPTION

The most plausible model for those who think that the goals of conservation biology must be more inclusive is a third utilitarian theory. This is the idea of option value, which links utility much more closely to diversity. Option value is a bet-hedging or insurance concept that conservation biology has borrowed from economics. The justification for thinking that option value is important rests on two plausible ideas. One is that species (or for that matter ecosystems) that are not of value to us at present may become valuable at some later time. In the more concrete language of economics, option value is the additional amount a person would pay for some amenity over and above its current value in consumption to maintain the option of having that amenity available for the future, given that the future availability of the amenity (its supply) is uncertain (van Kooten and Bulte 2000, 295). The second idea is that, as our knowledge improves (and as our circumstances change) we will come to discover new ways in which species can be valuable. Technically, this gives rise to "quasi-option value," that is the value of preserving options, given the expectation of growth in knowledge (Arrow and Fisher 1974). Following common practice in environmental ethics, we shall use the phrase "option value" to cover both types of value.

The crucial point about option value is that it makes diversity valuable. As we do not know in advance which species will prove to be important, we should try to conserve as rich and representative a sample as possible. As Daniel Faith notes, option value "links variation and value" (Faith 2003). So option value values unremarkable species and other aspects of biodiversity so long as, like species, they cannot be restored once they are gone. For example, there are many millions of beetle species (and likely to be many millions as yet undiscovered). Most represent very little demand value. Few are economically important (and some of those are important only as pests). While some provide important ecosystem services, very likely, most do so redundantly. They provide much the same ecosystem services as large numbers of other species. Beetles do not qualify as charismatic megafauna (many zoos exhibit no beetles at all despite the fact that they are easily the most speciose group within the animal kingdom). But for all this lack of notoriety, beetles do form many distinct species, each with their own unique mix of traits. Option value provides a justification for the preservation of these differences given that we might discover some of them to be of great importance.

So option value potentially applies to a broad group of species. A similar argument goes through for other aspects of biodiversity. There may be no significant demand value for a wetland now. But once drained and covered with housing, it has gone forever. No change of mind will then be possible. For these reasons, we judge it the best candidate ethical basis for a scientifically analyzable notion of biodiversity as a goal for conservation biology. In the sections that follow, we will argue that option value is also a de facto political and legal justification for much current conservation effort. We will also seek to answer two fundamental questions: If option value does give us reason to conserve species and ecosystems, how strong is the reason it provides? And what kind of biodiversity should we maximize?

Despite our optimism above, the task ahead of us is difficult. The option value model suffers from the same aggregation problem as every other version of utilitarianism: the option value of a given biological system (local population, species, multispecies community, or ecosystem) will be very different for different agents. Moreover, option value, understood one way, seems to be ubiquitous. If objects have an option value just in virtue of being useful in some imaginable future contingency, everything has option value, perhaps even identical option value. But if everything has option value, we cannot use its distribution to prioritize, to invest resources in one conservation project rather than another. If we choose to hedge our bets against any possibility whatsoever, then any morphological, developmental, evolutionary, genetic, behavioral, or ecological feature of any individual, species, or assemblage of species could prove valuable under some circumstances. That yields the useless goal of preserving biodiversity in all possible respects.

The solution is to focus not on mere possibilities but on probabilities. After all, many people successfully hedge against future contingencies (skiers pack chains "just in case," companies hedge against currency fluctuations, and so forth). This harnessing of option value is rational because it focuses on probabilities rather than on possibilities. The person who packs chains going to the beach in summer is right in thinking it possible that they will meet snow, but their decision is irrational because it is more probable they will meet sun and heat.

There is a further problem with the blanket assumption that biological traits might turn out to be good for something. If we are really ignorant of what the future holds, they might as easily turn out to be detrimental rather than beneficial. Elliot Sober's argument in his *Philosophical Problems for Environmentalism* attacks option value for turning ignorance of value into a reason for action. If conservation biologists are completely ignorant of the value of species, then they cannot make rational decisions either for or against their preservation (1986, 175). But we doubt the cogency of Sober's argument. It suggests that option-value arguments presuppose complete ignorance about the future benefits of biodiversity. If that were true, option value would be no guide to action. But, except in the abstract realm of thought experiment, it is not true. In the next three sections, we argue that relatively limited information about the taxonomy and ecology of threatened species tells us a surprising amount about their likelihood as sources of option value.

So the option-value approach to conservation biology depends on our being ignorant, but not too ignorant. Since we lack full knowledge about the future, we are wise to hedge our bets, insuring against unpleasant surprises. But we need to be knowledgeable enough to ignore very remote possibilities, to invest only a little against somewhat less remote possibilities, and to take serious measures to protect against more likely dangers.[7]

Importantly, one aspect of the world about which we are ignorant is our own future preferences. Here, the option value approach connects to the transformative value approach. Both of us are Australasians, and the ecologies of both Australia and New Zealand have been profoundly altered for the worse by deliberately introduced organisms. Some of these were just plain ecological mistakes—the cane toad is a failed biological control. But many of these alterations reflect a profound change in preference, in aesthetic sensibilities. Swamps are now wetlands; jungles are now rainforests. These rechristenings are reflections of changes in us as much as changes in our understanding of the biological world. A century ago, so-called acclimatization societies flourished in both of our countries. These had the goal of making Australasian ecosystems more like European ones. Most contemporary Australasians think that these sensibilities, sensibilities that motivated this undervaluation of the endemic biological world, were bizarre and wrongheaded. So one important source of option value is our insuring against changes in what we ourselves want and value.

One class of option value arguments will become less important as we improve our ability to predict the response of our biological environment to changes and interventions. The future will become more scrutable, and we will have less need to hedge our bets against unforeseen contingencies. God has no need for insurance policies. But improving our ecological understanding will do nothing to cure our ignorance of our own future preferences. Taking precautions to accommodate changes in our own desires will continue to be an important source of option value. We shall illustrate these issues by exploring the actual deployment of option value and to look at ways in which we have already discovered apparently unremarkable species to be importantly valuable. We shall look at three cases, each centering on a different aspect of option value.

8.5 APPLYING OPTION VALUE: CASE 1, PHYLOGENY

We said above that counting species is the most common means of assessing biodiversity. Species richness is a decent surrogate for phenotype disparity. For example, it is likely that species-rich communities are more stable in the face of disturbance than species-poor ones because species-rich ones have a wider range of phenotypes from which to meet the demands imposed by temporal and spatial variability. But, as we have argued, species richness also does capture a core component of biodiversity. Let's see how this plays out in an explicitly conservation biology setting, confining our discussion to sexually reproducing organisms, and to reproductive isolation. Option value explains the importance to us of reproductive isolation, via the link between isolation and evolutionary potential. While noting Mary Jane West-Eberhard's reservations, we have cautiously endorsed Douglas Futuyma's model of the link between speciation and phenotype divergence. Speciation allows daughter species to diverge radically in morphology, physiology, ecology, and behavior from their stem. For these reasons many people think of option value as mandating the preservation of species. We should deplore the extinction of any species because every species represents a new and potentially important trajectory in a space of evolutionary possibility. Most adaptive radiations began, in all probability, with a stem species that would have seemed only modestly different from their parent and sibling species. Evolutionary response can be rapid, so in framing option value in evolutionary terms, we need not be envisaging time scales of many thousands of years. But we are, it is true, presuming a multigenerational perspective on option value: conservation that depends on evolutionary bet hedging presumes that it's rational for us to insure against disasters that would impact future generations rather than our own.

Given that multigenerational perspective, species appear as natural loci of option value. But this leaves us facing a group of very important questions:

- How much option value is represented by the fact of speciation?
- How much conservation effort does speciation therefore justify?

- Do all speciation events represent the same amount of option value?
- Do some evolutionary trajectories represent more option value than others?

Although speciation events are undeniably important, that fact alone does not imply that species richness is the only good metric for biodiversity in conservation biology. Species counting should not rest on the assumption that all species represent equal amounts of biodiversity and that they are therefore of equal conservation value. In chapter 5, we explored the idea that species differ in evolutionary potential in virtue of differences in their population structure and their developmental biology. Moreover, species explore their evolutionary potential from their current location in morphospace, and so, even discounting intrinsic differences in evolutionary plasticity, the phenotypic divergence within a group of related species is important to the space of possibility to which they have access. We shall see that this idea is central to the importance of phylogenetic distance, to which we now turn.

As we noted in the last section, option value seems most important when the future is translucent rather than opaque or transparent. If we have complete information about our future, we need no insurance. If we have no information about the future, we cannot rationally hedge our bets, because we cannot spend limited insurance resources in any discriminating way. Given that we have some limited but imperfect information, what is the rational way to maximize our future options? Daniel Faith and a group of like-minded systematicists have linked option-value considerations to the idea that we should conserve as representative a sample of evolutionary history as possible. We should maximize the phylogenetic distinctiveness of the biota we conserve. We shall discuss phylogenetic distinctiveness in some detail, but (to borrow an example from 7.2) the intuitive idea is that a sample of two species of the genus *Ranunculus* is less phylogenetically distinctive than a sample of one species of *Ranunculus* and another species from a different genus in the same family (Ranunculaceae), because the two species in the second sample are more distantly related than those in the first sample, and hence they represent a larger and deeper chunk of the tree of life. Faith argues that this is an important feature of samples because "we do not know which traits will be of value in the future." We should therefore seek to "maximize representation among all of them" (Faith 2002, 250).

A recent study of floral diversity in South Africa suggests that maximizing phylogenetic distinctiveness[8] can lead to different conservation decisions than maximizing species richness, and that maximizing distinctiveness maximizes option value. Félix Forest and his colleagues argue that if we just maximized species richness, we would concentrate our conservation effort in the west of the Cape of South Africa, which is species rich as a result of a series of rapid radiations. But these radiations are very recent, and though there is a large species count, these are young, closely related species. In the east, the cape lineages are mingled with lineages that originated in a different biogeographic region. While we would maximize species number by concentrating on western reserves, a mix of east and west maximizes phylogenetic distinctiveness. Moreover, Forest and his colleagues argue that phylogenetic distinctiveness maximizes option value, for if we survey the past discoveries of economically useful plant species in the lineages in question, we find that they are scattered through the tree. Had we been making tough conservation decisions in 1900, maximizing phylogenetic distinctiveness would have given us our best chance of preserving the species that turned out to be useful (Forest et al. 2007; Mooers 2007).

The idea in play is that species represent option value because they are unique and potentially distinctive evolutionary trajectories. This recognizes speciation as a profoundly important process in the production of biological diversity. Among the many factors that influence the extent to which two species differ from each other are the length of time that the two species have been genetically isolated and the number of speciation events that have occurred since the existence of a common ancestral species. We shall call the conjunction of these two factors phylogenetic distance. One idea is that the phylogenetic distance between two species is roughly proportional to the amount of option value that they represent as an assemblage. As higher taxonomic richness within an assemblage is correlated with phylogenetic distance it will often be a good (albeit approximate) indicator of option value.

However, we have not yet addressed the problem of quantifying phylogenetic distance for option value. How much more is present in a small assemblage containing mammals, mollusks, and reptiles than in one composed only of primates? It is just this question that led to the development of "taxonomic distinctness" as a measure of biodiversity (Vane-Wright et al. 1991). The basic idea is to think of phylogenetic value as attaching to clades rather than to particular species. Each of these clades, when they are sisters of one another, is then assumed to contribute equally to the biodiversity of the system (and so in the current context we might further assume that each is assumed to constitute an equal amount of option value). An obvious way this can be achieved is to think of sister groups (those formed from a single speciation event) as representing equal amounts of biodiversity. The amount of biodiversity represented by each species is expressed as a weighting. An example of the way in which such a strategy works is given in figure 8.1.

The effect of this strategy is to make the products of older speciation events much more valuable than those of very recent events. It maintains the assumption that all speciation events are of equal importance. While this assumption is in line with cladistic views about the impossibility of privileging particular phenotypic traits, we think it is probably a mistake, both in practice and in theory. In practice it is problematic because the phenotypic distance between two taxa is much easier to assess than the speciation difference. There is, for example, a wide range of views about the speciation distance between humans and the two chimp species. We think it is also a theoretical mistake if option value is linked to maximizing the capacity of the surviving biota to respond to unforeseen contingencies on both ecological and evolutionary time scales. We think the phenotypic spread of the biota is the crucial dimension for buffering biota against disturbance. There is a helpful table of ecosystem services in a recent review of biodiversity loss and its consequences (Díaz et al. 2006) (table 8.1). Sandra Díaz and her colleagues argue that diversity stabilizes the delivery of all of these services, but in most cases, the diversity in question[9] is relevant phenotypic diversity ("functional diversity," in their terminology). This is what buffers ecosystem processes against disturbances.

Local morphospaces allow us to represent phenotypic diversity directly in a principled and tractable way, so it is not necessary to

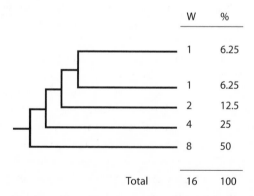

FIGURE 8.1. Equal weighting for sister groups. The cladogram represents speciation events with the ancestral species on the left. The species represented by the lowest horizontal line is the sister taxon to the clade containing all four other species, so its weighting matches the sum of all the other weightings. After Vane-Wright et al. (1990).

use phylogenetic distinctiveness as a proxy for phenotypic diversity. Similar considerations apply to evolution. As we noted in chapter 1, the living fossil phenomenon shows that privileging ancient splits, species like the platypus and Tasmanian devil, species whose most recent common ancestor with other living species lived a very long time ago, may not maximize evolutionary potential. These lone survivors of their lineages may be lone survivors because of a loss of evolutionary plasticity in their lineage. Moreover, all else equal, the space three taxa (for example) can explore expands with increasing distance in their initial positions. The phenotypic distinctiveness within an assemblage, then, is crucial to assessing the evolutionary potential of that assemblage, and, speciation distance is at best a proxy for phenotype difference. We share the cladistic suspicion of an overall phenotype space or morphospace. But as we argued in chapter 5, local morphospaces anchored around real lineages do permit us to make principled choices of dimensions and starting positions. In our view, a local morphospace is a better tool for representing differences in evolutionary potential between assemblages than this purely cladistic representation.

This approach does not put a dollar value on species.[10] Nor does it tell us how much conservation effort ought to be expended in the conservation of any particular species or assemblage of species. But it does tell us about the value of species and groups of species relative to one another. As it stands, though, it probably overestimates the option-value importance of distinctiveness, and underestimates the value of species richness. The number of points from which a space can be explored is as important as the average distance between those initial points. In a paper titled "What to Protect—Systematics and the Agony of Choice," Richard Vane-Wright and colleagues (1991) note that the

TABLE 8.1: Ecosystem Services Stabilized by Diversity

Quantity of Useful plant biomass production
Stability of plant biomass production
Preservation of soil fertility
Regulation of water supply
Pollination services
Resisting the invasion of harmful species
Control of agricultural pests
Climate regulation
Carbon storage
Buffering the impact of storms and similar disturbances

Source: From Díaz et al. (2006)

problem for the "equal weights for sister groups" strategy is that taxonomic rank overwhelms species richness. On this account the two
known species of tuatara have equal weight to the 6,800 species of
snakes, lizards, and amphisbaenians that make up their sister group.
If we think of this in terms of conservation effort, it leads us to a surprising conclusion. Assume that conservation organizations combine
to spend a million dollars on the conservation of the tuatara annually.
Then, by the "equal weighting for sister groups" algorithm, the total
annual conservation worth of each of the 6,800 species of snakes, lizards, and amphisbaenians that make up their sister group would be
about $147.

Of course we could (as always) do the mathematics differently, and
this is in fact the conclusion that Vane-Wright and his colleagues adopt.
They accept that species of tuatara represent more biodiversity than the
average species, but they think that what matters about tuatara is, not
the size of their sister group, but rather the fact that they are a member
of a very old taxonomic group that has very few species. As a solution to
this problem the authors propose their version of a taxonomic distinctness measurement. The aim of this algorithm is to pick out species with
few close relations by giving greater weighting to those species that are
members of fewer clades. As only extant species are taken into account,
species such as the tuatara that have a large number of extinct close
relatives still achieve a high weighting on this measure. An example of
the application of this strategy is given in Box 8.1.

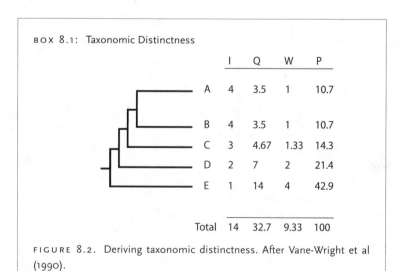

BOX 8.1: Taxonomic Distinctness

	I	Q	W	P
A	4	3.5	1	10.7
B	4	3.5	1	10.7
C	3	4.67	1.33	14.3
D	2	7	2	21.4
E	1	14	4	42.9
Total	14	32.7	9.33	100

FIGURE 8.2. Deriving taxonomic distinctness. After Vane-Wright et al
(1990).

The Vane-Wright strategy assigns an information content to each particular clade. It is measured by counting the number of statements about group membership that one is able to make about each taxon. In this example column I indicates the number of clades to which each species belongs within the system (the basic measure of taxonomic information). Column Q is the quotient of the total taxonomic information for the system divided by the information score for each taxon. Column W is the standardized weight (each Q value is divided by the lowest Q value). Column P gives the percentage contribution of each terminal taxon to the total diversity of the system.

Taxonomic distinctness seeks to leaven phylogenetic distance with a dash of species richness; notice though that there is still no attempt to represent phenotype distinctiveness explicitly, and we continue to think that this overlooks the important difference between overall and local morphospaces. Both these algorithms tell us that, viewed phylogenetically, there is an option value gulf between the phylogenetic rarities and the members of large and bushy clades to which the great majority of species belong. We have imperfect knowledge of threats and opportunities the world will bring to us, and we have imperfect knowledge of how our own preferences will change over time. A diverse, adaptable, evolutionarily plastic biosphere is like individual health. It is a fuel for success for our projects, both collective and individual. Such a biosphere is not a foundation for every project (no more than health is for individual projects). But it is for many, including many we cannot now anticipate. And so it is reasonable to invest in the preservation of the diversity of that biosphere. But species do not contribute equally to the existence of diverse, adaptable, evolutionarily plastic life. Recently evolved members of a species-rich lineage, like the snail darter, contribute less than the phylogenetically distinctive kagu. Even though we have assumed that each speciation is equally important as a potential producer of biological diversity on the basis of phylogeny, most species constitute very little option value, while a few score very highly indeed. So to the extent that phylogenetic distinctiveness captures the differences that may be important, but differences whose importance we cannot yet recognize, our conclusions are not democratic. Of course, it is possible that a novel, unnoticed mutation in the snail darter has put it on an evolutionary trajectory that will (if we can only recognize it) make it of enormous consequence for our own future projects. But this is a possibility akin to that we would recognize by taking snorkeling gear on a skiing holiday. We move now to two more specific proposals that link biodiversity to option value: bioprospecting and the provision of ecosystem services, an issue on which we have already touched.

8.6 APPLYING OPTION VALUE: CASE 2 BIOPROSPECTING

In this section we test the phylogenetic assumptions of the previous section against more practical applications of the idea of option value. Conveniently, we have a large and intriguing data set close at hand. The United Nations Convention on Biological Diversity is a detailed statement of the responsibilities of signatory countries toward their own biodiversity, but it also serves a commercial purpose. That purpose is closely bound up with the idea of option value.

A central problem for conservation biology has always been the fact that most of the ecosystems we want to save exist in the countries that have little money available for conservation. In poor countries serious endemic diseases, high infant mortality, and limited infrastructure inevitably seem more pressing than conservation. The elegant solution to this problem is set out in Article 1 of the Convention.

ARTICLE 1. OBJECTIVES

The objectives of this Convention, to be pursued in accordance with its relevant provisions, are the conservation of biological diversity, the sustainable use of its components and the fair and equitable sharing of the benefits arising out of the utilization of genetic resources, including by appropriate access to genetic resources and by appropriate transfer of relevant technologies, taking into account all rights over those resources and to technologies, and by appropriate funding.

This is an agreement to share the commercial spoils resulting from the cataloguing and conservation of biodiversity. The UN Convention exhorts poor countries to engage in conservation by promoting the idea that biodiversity is commercially valuable. One such value is bioprospecting; the collection and assessment of biological samples for economic purposes (for example, medicines, crops, and industrial products). The Convention endorses "the concept of nations holding property rights to their indigenous species" (Macilwain 1998, 537). This allows poor countries to negotiate, particularly with large pharmaceutical companies, for the exploration and exploitation of their "genetic resources." Companies can be charged for bioprospecting licenses, and as new commercially valuable substances are found, poor countries can be recompensed for their use. This in turn gives those countries reason to maintain their stock of biological diversity.

This is clearly an example of option value employed as a means of tying conservation to economic gain. It once seemed very successful. As Colin Macilwain puts it:

> The developing countries began to prepare for a gold rush of prospecting scientists from the United States and Europe. Their environmental ministers addressed the issue and made uncompromising public declarations of their readiness to strike a hard bargain—did everything, in fact, short of opening bars and brothels for the anticipated flood of bioprospectors. (1998, 535)

In the late 1980s and early 1990s the prospects looked bright indeed. It is widely recognized that many of our most important medicines have biological origins, for example, morphine, aspirin, and the polyketide antibiotics such as penicillin and streptomycin. A survey of drug discovery between 1981 and 2002 found that almost two-thirds of anticancer agents being investigated as drug candidates were derived from natural products (Newman et al. 2003). Less widely understood is the fact that known species of animals, bacteria, and particularly plants contain an extraordinary number of unique biochemical compounds. On the face of it then, biological specimens look like a promising source for biologically active compounds that might make their way into pharmaceuticals. As drug companies regularly spend half a billion dollars getting a drug to market, there were many predictions of positive conservation outcomes and of much-needed redistribution of wealth from prosperous temperate countries to the impoverished tropics. David Pearce and Seema Puroshothaman (1995) estimated that Organisation for Economic Co-Operation and Development (OECD) countries might suffer an annual loss of £25 billion if 60,000 threatened species were actually lost as a medicinal resource.

Despite early optimism, bioprospecting has fallen on hard times. Recent academic work has abandoned the breathless predictions of economic gains in favor of sober analyses of what went wrong (for example, Macilwain 1998; Simpson and Sedjo 2004; Craft and Simpson 2001; Barrett and Lybbert 2000; and Firn 2003). Large pharmaceutical companies such as Monsanto and Bristol Myers Squibb have shut down their natural products divisions entirely (Dalton 2004). Others have scaled down their natural product screening programs (Cordell 2000). According to these analyses, the earlier predictions overestimated the option value of unexplored biodiversity. While pharmaceutical companies spend a great deal getting new drugs to market, this does not imply that they will spend large amounts of money on licensing the "genetic property" of third world nations. Even without such fees bioprospecting is a very expensive business. Organically sourced compounds are difficult and therefore expensive to obtain. Along with the invention of the idea of property rights attaching to indigenous species came the invention of the idea of biopiracy (Gómez-Pompa 2004), the act of illicitly diverting the genetic resources for economic gain. So jumpy have some countries become that the charge has even been leveled at academic biologists who have no commercial intentions. Bioprospectors must therefore be very careful to document their work in ways that are acceptable to a sometimes large number of local officials.

As with any prospecting activity, there are many failures for every success. The vast majority of organic compounds collected show no useful biological activity. About 1 in 1,000 shows activity. About 1 in 250,000 yields a drug (Firn 2003, 210). As with all pharmaceutical development, the discovery that a compound is biologically active is just the beginning of a long process to establish that the resulting drug will be safe, clinically useful, effectively patented, marketed, extracted, synthesized, or produced by fermentation economically on an industrial scale. Just as with prospecting for fossil fuels, the whole industry is in competition with viable alternatives. Biochemists have become much more adept at producing synthetic compounds to test for biological activity. Furthermore, natural selection does not produce chemicals in the same way that industrial chemists do. It uses enzymes to produce large complex molecules. We use the brute force of chemical reactions. Thus it is much easier for us to synthesize large amounts of compounds that have been produced artificially. This means that naturally occurring chemicals usually have to be sourced from organisms that are easy to farm. It is thus no surprise that many of our most successful drugs come from microorganisms.

The declining fortunes of bioprospecting have led some to conclude that countries' biological resources are simply not as valuable as we currently assume (Simpson and Sedjo 2004). But even if bioprospecting had turned out to be a spectacular success, this would still not imply that the option value attaching to single species is sufficiently large to warrant serious expenditure for their conservation:

> The value to private researchers of the "marginal species" is likely to be small.... If there are many species that can serve as potential sources of new products, the probability of discovery among any species chosen at random must either be so high that two or more species are likely to contain the same chemical lead, or so low that none is likely to contain the lead. In either case, the expected value of having an additional species must be negligible. (Craft and Simpson 2001)

The news is not all bad. As noted above, a major problem for organically derived pharmaceuticals and agrochemicals is our limited ability to synthesize chemicals constructed naturally by enzymatic activity. This may severely hamper commercialization if the compound in question only occurs naturally at extremely low concentrations or the organism that produces it is very difficult to farm. However, recent work has explored the possibility of enhancing the chemical diversity of an organism by adding to it a gene coding for alien enzymatic activity (Firn and Jones 2000, 214). Such genes can be transferred between very different organisms (for example, from mammals to microbes). In theory this would allow us to "grow" chemicals sourced from organisms that would ordinarily not produce them in commercial quantities.

It is hard to say what effect such laboratory-based bioprospecting would have on the arguments advanced here. But it seems unlikely that it would alter the underlying economic argument put forward by Amy Craft and R. David Simpson. Even if it did, bioprospecting option value will weight phylogenetically distinctive species much more heavily than those from speciose clades. Bioactive chemistry is just a special case of evolutionary potential. As with our exploration of phylogenetic diversity, the story of bioprospecting tells us that, while all species are potential sources of valuable chemicals, this does not mean that all species should be seen as sufficiently valuable to warrant costly conservation measures. Pharmaceutical and agrochemical investment is decreasing and it was only ever a tiny proportion of total research and development budgets. This indicates that the "biochemical" option value of most species is very small. That may not matter. We do not have to save species one by one. Species exist as populations in ecological systems. If we protect those systems, we will save many of the species in them. Not all though. As we noted in chapter 6, there is no reason to believe that local communities are typically highly regulated. They are in flux, and the extinction of local populations is not infrequent (especially of small populations). Local populations can be reestablished by migration, but that is much less likely if the species exists in low numbers. Moreover, conservation in third world countries often involves making "damaged" ecosystems available for logging in return for the preservation of "pristine" habitats that will harbor viable populations of threatened species.

8.7 APPLYING OPTION VALUE: CASE 3, ECOLOGICAL OPTION VALUE

Ecosystems are highly interconnected, hence the common fear that species loss may lead to widespread ecosystem breakdown. Key ecosystems have very high demand value. Maintaining the health of, for example, major river catchments is of vast agricultural consequence. As we have seen in discussing the option value of phylogenetic distinctiveness, there is a case to be made that biodiversity is important because it makes the provision of key ecosystem services more reliable. In its simplest form, the argument goes something like this.

1. All species depend on other species via food webs, nutrient cycles, and phenomena such as niche construction.
2. Species can be driven to extinction via the disappearance of other species on which they depend.

Therefore:

3. Removal of any species from an ecosystem risks a domino effect, leading to wholesale species loss and ecosystem breakdown.

The claim is seldom put in such stark terms, but the idea underpins many important arguments in conservation ethics. A well-known example is Paul and Anne Ehrlich's (1981) claim that stressing ecosystems to the point that species are caused to go extinct is analogous to "popping" rivets out of an airplane in flight. Initially this activity has little effect. But at some crucial point the results are calamitous; a wing falls off. Given that this rivet popping (species extinction) has been going on for some time, we would be very foolish to ignore the threat of breakdown.

This line of thought is often coupled with ideas about redundancy and its limits (Walker 1992; 1995). The idea here is that many functional groups within ecosystems contain real but not limitless redundancy. So Paul Ehrlich (in a paper coauthored with Brian Walker, 1998) developed a version of the initial argument that turns on something akin to option value. Ehrlich accepts redundancy, but warns us that:

> A "redundant" species in a functional group that is exterminated today might well be the only species in the group that is able to adapt to new environmental conditions imposed on the ecosystem. (Ehrlich and Walker 1998, 387)

As we saw in 6.4, an important line of investigation suggests that redundancy buffers systems against change. Since we do not know which changes will challenge systems in the future, we should conserve redundancy. In doing so, we conserve unobtrusive species that may well come to play ecologically pivotal roles.

While the idea that redundancy buffers systems against disturbance very likely captures an important truth about ecological systems, we doubt that this shows that most species do play, or are likely to play, crucial roles in delivering ecosystem services. In most ecosystems a very small proportion of species have very high interactivity (they are either keystones or dominant species) and we know what sort of interactions are typical of such species. These include mutualisms such as pollination and seed dispersal (Soulé 2003, 1239). Effective predation is another typical keystone interaction, preventing overbrowsing and resultant simplification and even destruction of ecosystems. A typical example from the United States is overbrowsing of forests by native ungulates, including white-tailed deer (*Odocoileus virginianus*) and elk (*Cervus canadensis*) due to the loss of native carnivores such as the eastern timber wolf (*Canis lupus lycaon*). Niche construction by ecosystem engineers such as beaver (*Castor canadensis*) (Naiman et al. 1986) and elephants (*Loxodonta africana*) (Owen-Smith 1988) is another common keystone interaction. These strong interactions are not dotted randomly through phylogeny. They are more common in some taxa than others. For example, keystone species are often mammals (Soulé et al. 2003, 1244); indeed Geerat Vermeij has argued that there is a systematic tendency for species composed of organisms that have high metabolic demands (as mammals do) to play a disproportionate role in structuring biological systems (Vermeij 1999).

These considerations suggest that there is a class of vulnerable species that deserve conservation investment, for their extinction is likely to have large but unpredictable effects on ecosystem function. Marcel Cardillo and his colleagues have shown that large-bodied mammals with slow reproductive rates are especially vulnerable to human-caused environmental change. These animals—large herbivores and high-trophic level carnivores—are likely to have keystone effects, and so their loss might well be very serious (Cardillo et al. 2005; Cardillo et al. 2006). But considerations of this kind do not export to snail darter–style cases—restricted range variants of widespread ancestral stocks. The crucial issues here, as in the previous two cases, have to do with probabilities rather than possibilities. Species are not of equal importance to ecosystem function. Ecosystems degrade in predictable ways. These regularities have been called "community disassembly" rules (Worm and Duffy 2003). As we have noted, a good example is extinction by trophic level. Higher-level consumers are less diverse, less abundant, and under stronger anthropogenic pressure than those below them. Thus they face greater risk of extinction, but perversely they are often also important consumer keystone species (Duffy 2002). No rules in ecology are hard and fast, but regularities such as these give us good, if probabilistic, advice about the ecological value of groups of species in ecosystems under threat.

Species and species assemblages do have ecological option value. But it is rather more rare than the rivet-popping argument suggests, and it is very unevenly distributed. Again, the lion's share belongs to a small number of species. Practically, this underpins a strong case for ecological triage. If our concern is hedging against the collapse of crucial ecosystems, we should indeed be prepared to invest in the conservation of the basic structure of the food web, the conservation of important ecological engineers, the retention of some redundancy in suites of pollinators, and the like. Prudence requires us to treat communities and ecosystems as organized systems with crucial components whose continued operation cannot be taken for granted in the face of disturbance. But even given all this, many species with small populations and narrow distributions are unlikely to be appropriate targets of investment in virtue of their ecological option value.

8.8 THE CONSERVATION CONSEQUENCES OF OPTION VALUE MODELS

Decision theorists sometimes make a distinction between risks and threats. A risk is a possible outcome whose effects would be harmful and whose probability can be estimated. A threat is an outcome that may be calamitous but whose probability cannot reasonably be estimated. The crucial difference is that risks are events for which it is reasonable to prepare and for which we can gauge an appropriate level of preparation effort. By contrast, threats are dangers for which we cannot reasonably prepare beyond low-level information gathering and occasional reassessment to check that the threats have not become genuine risks. So, for example, global warming is a risk. There is a good chance of it causing major harm and we can make reasonable assessments of the probability of particular consequences of global warming causing harm to human populations. Threats gradually grade into risks as information improves. Sometimes we have good information about the probability of future contingencies. We know, for example, that the probability of serious droughts in eastern Australia in the next decade is very high. Sometimes we have qualitative information ("quite likely," "very unlikely"); sometimes, perhaps, we have not even rough qualitative information. In discussing the idea of option value, we have suggested that we typically are able to make at least rough qualitative estimates of future values and contingencies. If conservation is action under uncertainty rather than action under risk (that is, if the relatively likelihoods of the different outcomes are truly inscrutable) then the appeal to option value would be of little help.

Since we think we do have some information about probabilities, we think option value is an important desideratum in public debate about the conservation fate of unremarkable species such as the snail darter. Indeed, given that such species appear to be of no important benefit and are not the subject of public affection, we think that option value is likely the only important desideratum in their conservation. Moreover, option value links investment in conservation to diversity by making representativeness important. That said, the analysis in this chapter suggests that many species do not have high enough option value to justify major expenditure on their conservation, and some of these will be restricted range species that will not be protected as a by-product of investment in well-buffered ecosystem services. It is one thing to suppose that endangered species and rare ecosystems have option value; it is quite another to show that they will typically have sufficient option value to make them worth a major conservation effort. If option value is the right model for making conservation decisions, as we have suggested, our conclusions are light green rather than dark green. The option value option shows that many species are of great value, but it does not show that all species, or all biological systems, have important value and ought to be saved.

9 *Concluding Remarks*

At a meeting of the Australasian Association of Paleontologists, one of us (Maclaurin) was asked the following question:

> How does our view of biodiversity provide advice to working scientists? We claim that biodiversity hypotheses must be precise and testable. Each must be tied to a process that produces diversity or that regulates its influence. And now the scientist says that he's in a quandary. Apart from providing him with some broad characteristics of the nature of such theories, we cannot give him a procedure for identifying the theory he needs. Thus, he finds himself no better off than when he started. He cannot make his theory more precise and we have succeeded only in producing a sense of theoretical disquiet.

Our answer to this question brings out both our ambition and its limits. The prospects for the scientific investigation of biodiversity are not as bleak as those suggested by our worried interlocutor. At some point in any scientific enterprise, initial observations, hunches, and background knowledge have to be distilled into hypotheses that then become the focus of the enterprise in question. The investigation of biological diversity is no different in this regard from any other scientific enterprise. We cannot provide him with an algorithm, but we maintain that, on a case-by-case basis, it is often possible to successfully interpret claims about biodiversity. In so doing, we identify particular dimensions of biodiversity with the inputs and outputs of biological processes. In the preceding chapters, we have tried to support that idea with worked examples.

Moreover, it is important to avoid the false dilemma of believing that either there is a single natural quantity of a system, its biodiversity, or that anything goes, and that biodiversity is in the eye of the beholder. We suspect that our questioner really was assuming that the goal of identifying biodiversity must be that of identifying a single magnitude important for each discipline or even for the life sciences as a whole. Identifying a single quantity is clearly an attractive goal. It would prevent us talking past one another. It would maximize the collection of useful data. It would present a clear and unified picture of the natural world, useful to those promoting action in the face of ecological threats. But can such a one-size-fits-all measure adequately depict biodiversity? We doubt it. Species richness, supplemented in various ways, is a good multipurpose measure of biodiversity, because many processes affect richness (so it's a signal of their action) and it is causally relevant to many outputs (so it is a key causal driver). But we did say: "supplemented in various ways for various purposes." So we supplement it phylogenetically if our interest is in the ecological processes that build a biota; genetically and demographically, if we are interested in the conservation biology of the species in the system; phenotypically, if our interest is in the ways richness buffers disturbance. And so on.

Our discussion of species richness illustrates, we think, the falseness of the dilemma above. Species richness does not measure the diversity of biological systems, but we can develop explanatorily powerful and broadly applicable measures of biodiversity, many of them based on, and extensions of, species richness. Indeed, we think this is the central and viable project of biological taxonomy. But even while agitating for broad consensus on a version of the evolutionary species concept, we recognize that such a species concept will not suit all theoretical and practical purposes. Nor is it applicable to all organisms.

In some enterprises, notably conservation biology, there are good methodological and political reasons for championing a one-size-fits-all concept of biodiversity. In defending pluralism as we do, it is important to recall the distinction between biodiversity and biodiversity surrogates. There are reasons to do with time, resources, and commensurability to prefer a one-size-fits-all measure or surrogate. Place prioritization algorithms (of the kind championed by Sarkar 2005) exist for a reason. Conservation policy makers must prioritize, and it is essential in institutional settings to have decision-making procedures that rely on measures that are transparent, intersubjective, and repeatable. We see the force in these reasons, but we maintain that whether or not they are ultimately persuasive is an empirical issue. Whether there is a single best measure of biodiversity depends upon the facts, not our institutional practices.

Political will, ease of measurement, and availability of data cannot trump the facts about the forms of diversity that have driven a system in the past, and that drive it now. How many dimensions of biodiversity are in play? Are some of the processes involved so powerful as to swamp others? Are any of the present dimensions of biodiversity a good proxy for all or at least almost all of the others? There can be no full and final answers to such questions, both because there are limits to our knowledge and because the causal drivers of biological systems themselves change over time. Moreover, there are concepts of biodiversity that have their home in policy and politics rather than the life sciences proper. Charismatic megafauna are charismatic, and they may often be important as keystone species. But they are not important because they are charismatic. Recent work by David Stokes (2007) on the psychology of the aesthetic evaluation of animal form makes clear that there is very little correlation between the importance of an organism for the ecosystem it inhabits and human evaluations of its aesthetic value. But despite the mismatch, neither characteristic can be ignored in the conservation of biodiversity, if only because many of those funding conservation see themselves as funding the conservation of the charismatic megafauna. Likewise, wilderness is an aesthetic or moral concept rather than a biological one, yet it has played an important role in conservation policy. Where policy and value meet biology, there are aspects of biodiversity that exist relative to human purposes and human values, and these are subject to change. They are features of our response to the biota rather than features of the biota itself. That said, the cases presented in this book have allowed us to develop some preliminary conclusions.

9.2 THE VARIETY OF DIVERSITIES

We have argued for a multidimensional view of biodiversity, though without (of course!) identifying all the dimensions, or specifying their relations one to another. We have done so mostly by tracking the fate of species richness as a core concept of biodiversity. Species richness is a core concept because (i) it is (relatively) theoretically precise; (ii) it is (relatively) easy to measure; (iii) it is (relatively) uncontroversial that the species richness, and species identity, of biological systems are important to the dynamics of those systems. Nothing is certain. It is possible that ecological function and ecosystem service are best understood by analyzing the relationships between the functional groups in ecosystems rather than by identifying specific species and tracking their relations. Even so, we agree with this consensus. But as we remarked in 9.1, it is also clear that richness has to be elaborated in different ways for different biological purposes.

Moreover, in exploring and defending concepts of biodiversity based on species richness and related measures, we certainly do not endorse a species equivalence principle. There is no biologically intelligible sense in which a mammal species and a mollusk species represent the same amount of biodiversity. A mollusk species might be the same kind of thing as a mammal species because it has been generated in the same way. But in discussing morphological and developmental spaces, we argued that while disparity and plasticity were important aspects of biodiversity, they had to be relativized to particular clades. There is no Global Morphospace or Library of Mendel in which mammals and mollusks can be meaningfully compared. We can compare one mollusk to another, but as species become increasingly phylogenetically distant, those comparisons become increasingly strained and arbitrary. So species richness measures between systems or across time are most meaningful when we compare richness of one clade with richness of that same clade. Tropical rainforests and coral reef systems are both species rich, but we learn very little in comparing reef richness with rainforest richness.

In claiming that processes underpinning dimensions of biodiversity differ greatly from one another, we do not mean to imply that they are independent of one another. Nobody now doubts that there are important causal links between evolution and development. Developmental mechanisms have themselves evolved: they reflect past interplays between ecology and chance. And development, by providing variation, makes further evolutionary change possible. Recent work suggests that development itself is less uniform than previously assumed. Gene expression is not uniform, for developmental plasticity is a precursor to adaptation. Even mutation is not isotropic, differing both within and between species. Accumulation of variation is subject to population structure, mating system, and environmental uniformity (sometimes engineered). Thus the evolution of development literature suggests that rather more dimensions of biodiversity than one might expect are ultimately hostage to the vicissitudes of current and past ecology.

In a number of places in this book, we have suggested that there is a fundamental distinction between those phenomena that are products of biodiversity and those that are sources of it. While biological diversity is studied for many purposes it is most often invoked as a tool for conservation, and here the distinction is most crucial and most problematic. Conservation biology must inevitably exploit those processes that give rise to diversity, but it is ultimately motivated by the great variety of consequences that biodiversity brings about. Even at this early stage it is clear that the aspects of biodiversity that underpin conservation efforts are very different from one another. Source questions are more tractable than consequence questions, because they are more straightforwardly empirical. It is true that the charismatic megafauna are as dependent on ecosystem services as any other group of organisms and equally true that the next aspirin-or-penicillin-size discovery might derive from something that is both visible to the naked eye and beautiful as well. Nonetheless, those aspects of biodiversity that are valued by humanity are unlikely to be captured by a single metric.

The problem of biological conservation is both empirically and philosophically hard. Thus we care about biodiversity as a source of pleasure, as a reservoir of invention, and as a force for the maintenance of global ecological stability. But these parameters are difficult to measure. We can monitor the successes of ecotourism and bioprospecting and we certainly witness the effects on ecosystem services that are caused by poaching, pollution, and habitat destruction. However, the positive outputs from biodiversity have proved difficult to translate into indexes for its measurement. While there is some data suggesting uniformity in aesthetic judgments by humans of individual morphologies, and hence the beginnings of an understanding of why charismatic fauna are charismatic, we are a long way from understanding the wider phenomenon of biodiversity as a source of human enjoyment. The values of landscape, wilderness, and sheer variety are clearly important to ecotourism, but how to measure systems in ways that capture those values remains a problem.

The jury is still out on the relationship between dimensions of biodiversity and the maintenance of ecological stability and productivity. While biodiversity surrogates can be good indicators of failing ecological health, they are at best imprecise markers of the changes in biodiversity that brought that ill health about. Measuring biodiversity as a possible source of yet-to-be-discovered scientific and commercial resources has proved even more difficult. While this elusive commodity is commonly assumed to be tracked by species richness, it is difficult to back up that assumption with either data or argument. Actual exploitation of biodiversity for commercial purposes has had a checkered history. Current economic analysis suggests that if there's gold in them there clades, there's not much of it and it is hard to find. We have suggested here that taxonomic diversity is a more plausible proxy for such diversity than is species richness.

9.3 SHOULD WE CONSERVE SPECIES?

We began with species in a conservation biology setting, and we will end there. First, a caution. We have focused on difficult cases, but there are many uncontroversial aspects of biodiversity that are demonstrably important in conservation settings. Loss of genetic diversity destroys potential adaptive pathways, making species ripe for extinction. Loss of trophic diversity disassembles communities at least as effectively as do fires or bulldozers. We depend for our existence on a host of ecosystem services, and they depend for their resilience on the buffering that redundancy brings. We have much yet to learn about the conservation consequences of many other aspects of biodiversity, but this is no reason to assume them to be unimportant.

Moreover, the conservation of species is linked to other aspects of biodiversity conservation by the use of species richness as a proxy measure. The use of proxies of biodiversity is essential, and measuring species richness is a good one. But to the extent that informational and resource constraints allow, we should calibrate and recalibrate our proxies. A good proxy for one aspect of diversity may not be a good proxy for others; butterfly richness is very likely a good proxy for plant diversity, but it may well not be a good proxy for plant disparity. It is important to calibrate our proxy measures, and calibration becomes more important the more we use a proxy and the greater the range of decisions we use the proxy in.

Leaving aside these practical issues, there can be no doubt that most people desire the conservation of the diversity of life on earth. Perhaps this wish bundles together dimensions of biodiversity that are strictly incommensurable. Certainly we do not as yet have a translation of this imperative into any simply applicable measure of biological difference, and perhaps we will never have one. Even so, the desire to conserve diversity will still be important politically and economically and it will still provide strong motivation for the conservation of regional biota, and for the conservation of those species that the public recognizes as distinctive and important. But the aim of conserving some of the species some of the time does not translate into a rational intent to preserve all of the species all of the time. We are not always blessed with local people with the resources to conserve their own local and valued biota. Similarly, there are many species that are not distinctive or that have not as yet caught the public's attention.

Conservation biology is so named because it has seemed plausible up to now that we might maintain many of the world's ecosystems roughly as they are, still with much the same complement of species. Many current models of global warming suggest that such an outcome may soon be rendered impossible. If the world climate changes radically, humanity will be faced with the need for ecological engineering just to keep pace. Species will have to be moved great distances. Lost keystones will have to be replaced with unfamiliar best guess alternatives. Sacrifices will have to be made. In such a world we would be faced with stark choices about how much effort to put into a democratic rescue of each and every species. Triage is likely to become more pressing, not less pressing.

Species richness does reflect a dimension of biodiversity, but it does not follow that every species warrants a serious conservation effort. Ultimately, the risk posed to humanity and to the wider biosphere by the extinction of any particular species, or indeed any particular clade, is an empirical issue. The trick for conservationists is to choose those that matter most (and to conserve as many as possible as side-effects of other conservation measures). Although speciation is essential for the production of adaptive variety, species counts may not always be a good marker of the presence of such variety. Species that have diverged very recently are unlikely to contribute to diversity as gauged by other, particularly physiological, dimensions of biodiversity. On average and over the long run, differentiation takes time. So we think if there is to be any general call for the conservation of biodiversity tout court on the basis of option value considerations, it should be for the conservation of phylogenetic and phenotypic diversity rather than mere species richness. We should sample as richly as possible the phenotypic and evolutionary resources of a biota, and in a world of limited resources, that might not involve maximizing species richness.

Notes

1. For a good discussion of current extinction rates, comparing them to mass extinctions of the past, see Sarkar (2005).

2. This was certainly the case in Britain (Sheail 1976), although the situation was somewhat better in the United States (Bocking 1997, 23).

3. For an insightful and wonderfully gossipy discussion of the twentieth controversies about this system, see Hull (1988). For a recent and in-depth analysis, one rather skeptical of the usual sharp distinction between species and other taxonomical ranks, see Ereshefsky (2001).

4. And hence it is not true of those parts of biology that only study living systems as complex physical and chemical systems; for example, parts of molecular biology and physiology.

5. Perhaps more accurately: the explanations offered by ideal morphology were increasingly out of step with nineteenth-century science; see, for example, Desmond (1982) and Rupke (1994).

6. "Character" is the technical term in taxonomy that refers to properties used in taxonomic investigations. A typical definition states "[a] character (e.g., eye colour) is an observable feature of a taxon that is variably exhibited in subdivisions called character states (e.g., red, blue, yellow)" (Cranston et al. 1991, 113).

7. The discipline was originally known as "numerical taxonomy," from the title of the book by Robert Sokal and Peter Sneath (1963). It was named "phenetics" by Ernst Mayr in 1965 (Hull 1988, 132), because after the rise of cladistics (described in the next section) phenetics seemed more distinctive for its emphasis on phenotype than for its emphasis on numerical methods.

8. While parsimony analysis is by far the most widely used method for a variety of practical and theoretical reasons, it is not the only type of analysis available to cladists. For a description of other types of analysis, see Quicke (1993, 52). For general introductions to this view of systematics, see Brooks and McLennan (1991; 2002) and Harvey and Pagel (1991).

9. For a nice discussion of gene trees, and of some of the limitations of using them as proxy for organism lineages, see Dawkins (2004, 44–55).

10. Although not always; sometimes the terms "guild" or "functional group" are used for a set of phylogenetically close populations, each of which has a similar ecological role; for example, populations of grain-eating rodents. See, for example, Ernest and Brown (2001).

11. There are general formal frameworks for representing biodiversity (and, indeed, diversity in general), but these require a prior and theoretically coherent set of choices that identify units, traits, and the relative importance of those traits. For a presentation and discussion of such a framework, see Puppe and Nehring (2002; 2004).

12. We pick this example because these lizards have been a vehicle for much study of the effects of competition and niche overlap. They are distributed in complex ways over many Caribbean islands, sometimes one to an island, sometimes more. See Roughgarden and Pacala (1989) and Roughgarden (1995).

13. Or perhaps more accurately, the genealogical relations between living taxa, for some cladists are skeptical about the possibility of incorporating taxa known only from fossils into a well-motivated phylogeny. On this, see Grantham (2004).

14. The distinction we have in mind is parallel to the one Elliott Sober draws between "source laws" and "consequence laws" (1984, 50–51). Source laws describe the origins of fitness differences; they explain why alleles are differentially fit. But most models in population genetics are consequence models; they predict the downstream consequences in a population of the frequency of different alleles, given their fitnesses and current relative frequency. But they say nothing about why alleles are more or less fit. Source laws explain why different alleles are differentially fit.

CHAPTER TWO

1. Paleobiological work on diversity nearly always measures it by numbers of higher taxa in a biota, often families. In part, this is due to limits on information we can access through the fossil record.

2. Some think this whole shift is misguided. Thus John Dupré in his "In Defence of Classification" argues that species should not be treated as the fundamental units of classification and not, therefore, as units of evolution (2001, 203). Likewise, Mishler and Donoghue (1982, 497) talk of "decoupling the basal taxonomic unit from notions of 'basic' evolutionary units."

3. In contrast (as we shall see in the next chapter), there are those who think the animal phyla are something like a natural kind, and hence there is a debate about the nature of a phylum.

4. In her 2003 volume, West-Eberhard questions the crucial premise, denying that local adaptation depends on gene combinations vulnerable to dilution effects, and arguing instead that they typically depend initially on mechanisms of adaptive plasticity. We return to this issue in chapter 5.

5. For a good recent statement of this perspective, see Eldredge (2003) and Thompson (2005). For a more cautious assessment of the extent and consequences of gene flow between distinct populations, see Morjan and Rieseberg (2004).

6. Stephen Jay Gould is as far from a neutral commentator on this issue as one could be. But chapter 9 of his massive *Structure of Evolutionary Theory* (Gould 2002) includes an extensive review of the empirical literature on stasis. In very recent work, Hendry (2007) and Estes and Arnold (2007) also accept that stasis is a widespread, pervasive phenomenon—rapid short-run evolutionary change does not typically sum to long-term evolutionary change—though they do not accept that the mechanism described here explains that pattern.

7. In doing so, he used a version of the cohesion concept.

CHAPTER THREE

1. Although some of the traditionally recognized mass extinctions may turn out to be periods of reduced speciation coupled with normal extinction rates; see Bambach et al. (2004).

2. Some phyla have very poor fossil records, so it is certainly not true that all the living phyla have deep and unmistakable fossil representatives. But none appear relatively late (say, in the Jurassic) and then have a good, continuous history once they have appeared. Nor, as far as we are aware, is there molecular clock data suggesting that some living phylum has split recently from another.

3. For the Chinese sites, see Hou et al. (2004). For Greenland, see Conway Morris et al. (1987) and Budd (1998). For the Proterozoic embryos, see Chen et al. (2000).

4. For overviews, see Smith and Peterson (2002), Bromham (2003), and Dawkins (2004).

5. For an excellent review of the early history of the bilaterians, and of the relevance of the flatworm debate to the important issue of the complexity of the early bilaterians, see Baguña and Riutort (2004) and Littlewood et al. (2004).

6. See, for example, the trees on pp. 24–25 of Fortey et al. (1996). There they discuss various phylogenetic representations of explosive hypotheses about the Cambrian radiation, and none are explicit about origins or branching order.

7. Davidson and Erwin seem to accept something like this view: "Critically, these kernels would have formed through the same processes of evolution as affect the other components, but once formed and operating to specify particular body parts, they would have become refractory to subsequent change" (Davidson and Erwin 2006, 799).

CHAPTER FOUR

1. While this example is extremely simple, size is often an ecologically and evolutionarily informative trait. For example, one response to predation (including human fishing) is selection for accelerated sexual maturity and hence reduced adult size. So variation in size can be a quite salient marker of variation in the intensity of predation.

2. Technically, shape is "the geometric information that remains when location, scale and rotational effects are filtered out from an object" (Kendall 1977).

3. McGhee (1999; 2006) is strongly in favor. Eble (2000) provides more cautious support. McGowan (2004) is opposed.

4. Fair game extinctions are selective in the Darwinian sense, while in wanton extinctions some kinds of organisms survive preferentially, but not because they are better adapted to their normal environments.

5. For a clear discussion of this problem, see Dennett (1995, 103–7).

6. It should be noted that this disagreement is partly about what constitutes a function. Lauder's definition specifies a level of kinematic precision that Plotnick and Baumiller see as overly restrictive given the general epistemic constraints of paleobiology.

7. Examples include Cambrian marine arthropods, Paleozoic gastropods, Paleozoic rostroconch mollusks, Paleozoic stenolaemate bryozoans, Paleozoic seeds, Cretaceous angiosperms—based on their pollen, Cenozoic ungulates, Carboniferous ammonoids, Paleozoic articulate brachiopods, Ordovician trilobites (but not Paleozoic trilobites as a whole), early–mid Paleozoic tracheophytes, Paleozoic crinoids, Mesozoic crinoids, Paleozoic blastozoans, and Cambrian Metazoa.

8. Counterexamples include early Jurassic ammonites, Paleozoic trilobites, and Paleozoic blastoids.

CHAPTER FIVE

1. This has led to a technical subfield of conservation biology concerned with "population viability analysis," modeling the time to extinction of populations in the face of stochastic fluctuation in recruitment and mortality (clearly, small populations are at much greater risk of unpredictable dips to zero) and using these techniques to estimate a "minimum viable population." For clear but skeptical explanations of this approach, see Caughley (1995) and Sarkar (2005).

2. See, for example, Daugherty et al. (1990).

3. Between-population variability can be measured in a number of different ways, depending on whether we are interested in differences of gene frequency as well as alleles found in one gene pool but not found at all in another: two gene pools may contain exactly the same alleles, but in different frequencies.

4. For the purposes of this chapter we shall go along with the idea that all biological inheritance is genetic inheritance. This is at best controversial. Eva Jablonka and Marion Lamb, for example, have argued that there are four inheritance channels. One is genetic inheritance. A second is epigenetic inheritance—mostly molecular mechanisms that influence whether and how a transmitted gene is expressed. A third is behavioral inheritance, and the fourth is cultural inheritance mediated by symbolic communication (Jablonka and Lamb 2005). The importance and extent of these mechanisms remain to be settled, so despite our suspicion that they are important (Sterelny 2001b; 2004), we shall set aside the role of nongenetic inheritance in diversity.

5. The classics in this rich field are: Lewontin (1985), Kauffman (1993; 1995), Dawkins (1996), Raff (1996), and Wagner and Altenberg (1996).

6. The other part is the nature of the adaptive landscape, for the existence of many local optima tends to damp down evolutionary change.

7. For example, Davidson and Erwin have appealed to this mechanism in their explanation of conservative features of gene regulatory networks (their kernels); see Davidson and Erwin (2006).

8. Although not usually at the origin of a clade, and hence the distinction between stem and crown taxa; stem taxa need not manifest the phylum-typical body plan.

9. Importantly, both Peter Wagner's and Günter Wagner's work include models of evolutionary change in modularity itself (see Wagner 1995a, 1995b; Wagner and Altenberg 1996).

10. For an alternative picture of the evolution of modular development see Dawkins (1996). He argues that there is high-level selection for lineages that have developmental systems that potentially enable them to generate a wider spread of variation.

CHAPTER SIX

1. See, for example, Gaston (2000).

2. For a good discussion of these issues, see Ward and Thornton (2000). They argue that early in their reassembly, Krakatau island communities do not show much path dependency. Predictable, good colonizers arrive early and everywhere. Order effects exemplify contingency only if the order of arrival is itself unpredictable. But communities then diverge as a result of chance differences in colonization. These differences have caused at least a significant delay in reaching of the equilibrium condition that seems to be dominated by *Dysoxylum* forest.

3. That is, within-patch richness versus between-patch richness; more on this in chapter 7.

4. This debate using this terminology began with an important paper by Jared Diamond (1975) on New Guinea bird faunas, in which he argued that competition for resources structured communities. Birds whose needs were too similar could not both be present, but nor would communities have, at equilibrium, underutilized resources either: species are densely packed into communities.

5. Jay Odenbaugh (2006) has introduced some useful terminology to describe different ensembles. "Gleasonian" communities contrast with "Hutchinsonian" and "Clementsian" communities. These community types are named for the ecologists who famously defended different views of communities. Gleason was an early American ecologist who argued that populations respond to their environments largely independently of one another. Hutchinsonian and Clementsian communities are distinguished by the strength of the interactions among their component populations. Clementsian communities, as Odenbaugh defines them, have strongly interacting components, whereas populations in a Hutchinsonian community may interact weakly.

6. For an insightful discussion of this complex of ideas, see Cooper (1993; 2001; 2003) and Pimm (1991).

7. We have presented this argument from persistence to regulation as if it were a single view. In reality, there is a family of views that accord the "balance of nature" a central role in regulating the internal organization of communities. In part, this

is the obvious point that stability comes in degrees. But it is also due to the fact that as Stuart Pimm and others have shown, the notion of stability is itself ambiguous; see Pimm (1991) and Lehman and Tilman (2000). There is a family of notions, and hence a family of ideas, about the extent to which communities are stable. But according to all these views, communities have a genuine organization. The organisms already present and interacting will make certain roles available and will foreclose others. For a particularly good discussion of the range of stability-conceptions that are lumped together, see pp. 118–21 of Sarkar (2005).

8. These are two of New Zealand's richer and more species-rich tidal mudflat communities—though nowhere near as rich as equivalent Australian systems.

9. Although there are not very many studies demonstrating this; for one, see Woodward (2002).

10. As we have noted, stability comes in many forms (Sarkar 2005, 118–21). The literature that has grown up around Tilman focuses on variance around a mean over time.

11. This data would have to be interpreted very cautiously. Paleoecological data does not record events on the same spatial and temporal scales as ecological studies on local communities.

12. This issue is fraught and controversial. There is a rich literature on food webs and the effect of the complexity and depth of food webs on stability. See, for example, Pomeroy (2001), Montoya et al. (2003), and Thebault and Loreau (2003).

13. The lottery effect raises subtle issues about causation as well as ecological mechanism. Drift has often been interpreted as a random change in populations over time, but recently, Patrick Forber and Ken Reisman have pointed out that on a manipulationist conception of causation, drift is a cause of population change. We can intervene on population size to change the importance of drift; we can manipulate the impact of drift on evolution. In small populations, it is more important. So drift is a lever we can use to influence evolutionary trajectories, and that makes it a cause. On a similar criterion, even if the lottery effect is the mechanism of the diversity-stability connection, species richness is the cause of stability. We can manipulate the impact of the lottery effect, and hence stability, by manipulating species richness. See Reisman and Forber (2005).

14. Levins and Lewontin made this point forcefully, and their ideas have been further developed by those working on niche construction and ecological engineering (Levins and Lewontin 1985). See also Jones et al. (1997) and Odling-Smee et al. (2003).

CHAPTER SEVEN

1. Certainly the most widely cited.

2. Surrogates are often samples: bird and butterfly species richness are often used as measurable surrogates of overall species richness, but they are also part of overall species richness.

3. Note though that there is considerable disagreement about the extent and nature of such correlations. In favor of strong correlation, see Gaston and Williams

(1993), Williams and Gaston (1994), and Balmford et al. (2000). Against, see Prance (1994) and Grelle (2002).

4. For a full discussion of the sort of features required in an indicator taxon, see Pearson (1995).

CHAPTER EIGHT

1. Those who want more information might consult one of many good general surveys of environmental ethics. Recent examples include Wenz (2001), Light and Rolston (2003), and Sarkar (2005).

2. For an example of such intuitive reasoning, see Ehrenfield (1981, 177).

3. It is not just those from outside the ethics community who have reservations about the worth of ethics based upon intrinsic value. There is also debate within the ethics community. See, for example, Agar (2001), and Sarkar (2005).

4. For a review of these issues, and a debate about the cogency of this approach to conservation biology, see *Ecological Economics* (vol. 25, no. 1, 1998), a special issue on this topic.

5. Medical ecosystem services do depend on specific species, for they depend on the specific bioactive compounds a species makes. But in a recent paper on the demand value of central South American forests (in particular, of Atlantic forest remnants in Paraguay), these were the least important of the ecosystem services. See Naido and Ricketts (2006).

6. Supposing, for example, that all agents who engage in rational reflection will converge on placing a high value on ecotourism and other natural history pursuits that depend directly on high levels of diversity.

7. As we see it, then, because option-value reasoning is tied to some future assessment, albeit rough, it does not depend on the decidedly controversial "precautionary principle," beloved of green politics. On that principle, see, for example, Sandin, (2005), or the very skeptical Sunstein (2005).

8. Forest and colleagues use a different measure of distinctiveness from that developed by Faith. They use molecular clock estimates to turn branch lengths in the phylogenetic trees they construct into estimates of the elapsed time since the last common ancestor of the species being compared. They add these times up to give a measure of the evolutionary history the samples represented.

9. In a few cases, they think that within-population genetic diversity is the crucial component of diversity, and water regulation and climate effects depend as well on the size and structure of the landscape through which the biota is spread.

10. Indeed, even if the value of diversity to human welfare were sharply defined, there would be no easy way to move from welfare value to a dollar value, for the dollar value depends not just on welfare effects but on income: how many dollars agents have to spend. Income is very unequally distributed, and Díaz and her colleagues point out that this has malign effects for diversity conservation, because the poor are the most dependent on ecosystem services and have the least resources for protecting those services.

References

Agar, N. 2001. *Life's intrinsic value: Science, ethics, and nature*. New York: Columbia Univ. Press.

Alroy, J. 2000. Understanding the dynamics of trends within evolving lineages. *Paleobiology* 26 (3): 319–29.

Andelman, S., and W. Fagan. 2000. Umbrellas and flagships: Efficient conservation surrogates or expensive mistakes? *Proceedings of the National Academy of Science* 97 (11): 5954–59.

Arnold, M. L., A. C. Bouck, et al. 2004. Verne Grant and Louisiana irises: Is there anything new under the sun? *New Phytologist* 161:143–50.

Arrow, K. J., and A. C. Fisher. 1974. Environmental preservation, uncertainty, and irreversibility. *Quarterly Journal of Economics* 88 (2): 312–19.

Arthur, W. 1997. *The origin of animal body plans: A study in evolutionary developmental biology*. Cambridge: Cambridge Univ. Press.

———. 2000. The concept of developmental reprogramming and the quest for an inclusive theory of evolutionary mechanisms. *Evolution and Development* 2 (1): 49–57.

———. 2001. Developmental drive: An important determinant of the direction of phenotypic evolution. *Evolution and Development* 3 (4): 271–78.

———. 2004. *Biased embryos and evolution*. Cambridge: Cambridge Univ. Press.

Babcock, L., W. Zhang, and S. Leslie. 2001. The Chengjiang biota: Record of early Cambrian diversification and clues to exceptional preservation of fossils. *GSA Today* 11 (February): 4–9.

Baguette, M. 2004. The classical metapopulation theory and the real, natural world: A critical appraisal. *Basic and Applied Ecology* 5:213–24.

Baguñá, J., and M. Riutort. 2004. The dawn of bilaterian animals: The case of the acoelomorph flatworms. *Bioessays* 26 (10): 1046–57.

Balmford, A., A. J. E. Lyon, et al. 2000. Testing the higher taxon approach to conservation planning in a megadiverse group: The macrofungi. *Biological Conservation* 93:209–17.

Bambach, R. K., A. H. Knoll, and S. C. Wang. 2004. Origination, extinction, and mass depletions of marine diversity. *Paleobiology* 30 (4): 522–42.

Barker, G. M. 2002. Phylogenetic diversity: A quantitative framework for measurement of priority and achievement in biodiversity conservation. *Biological Journal of the Linnean Society* 76 (2): 165–94.

Barrett, C. B., and T. J. Lybbert. 2000. Is bioprospecting a viable strategy for conserving tropical ecosystems? *Ecological Economics* 34:293–300.

Bell, G. 1989. *Sex and death in protozoa: The history of an obsession.* Cambridge: Cambridge Univ. Press.

Bengston, S. 2006. A ghost with a bite. *Nature* 442 (July 13): 146–47.

Bennett, K. D. 1997. *Evolution and ecology: The pace of life.* Cambridge: Cambridge Univ. Press.

Berlin, B. 1992. *Ethnobiological classification: Principles of categorization of plants and animals in traditional societies.* Princeton, NJ: Princeton Univ. Press.

Biles, J. A. 1994. GenJam: A genetic algorithm for generating jazz solos. *Proceedings of the 1994 International Computer Music Conference* (San Francisco): 131–34.

Blondel, J. 2003. Guilds or functional groups: Does it matter? *Oikos* 100:223–31.

Bocking, S. 1997. *Ecologists and environmental politics: A history of contemporary ecology.* New Haven, CT: Yale Univ. Press.

Bookstein, F. L. 1991. *Morphometric tools for landmark data.* New York: Cambridge Univ. Press.

Bottjer, D., J. W. Hagadorn, et al. 2000. The Cambrian substrate revolution. *GSA Today* 10 (9): 1–7.

Bowring, S., J. Grotzinger, et al. 1993. Calibrating rates of early Cambrian evolution. *Science* 261:1293–98.

Brandon, R. 1999. The units of selection revisited: The modules of selection. *Biology and Philosophy* 14:167–80.

Briggs, D. E. G., R. A. Fortey, and M. A. Wills. 1992a. Morphological disparity in the Cambrian. *Science* 256:1670–73.

———. 1992b. Reply to Foote & Gould and Lee. *Science* 258:1817–18.

Bromham, L. 2003. What can DNA tell us about the Cambrian explosion? *Integrative and Comparative Biology* 43:148–56.

———. 2006. Molecular dates for the Cambrian explosion: Is the light at the end of the tunnel an oncoming train? *Palaeontologia Electronica* 9 (1).

Brooks, D., and D. McLennan. 1991. *Phylogeny, ecology, and behavior: A research program in comparative biology.* Chicago: Univ. of Chicago Press.

———. 2002. *The nature of diversity: An evolutionary voyage of discovery.* Chicago: Univ. of Chicago Press.

Brooks, T. M., R. A. Mittermeier, et al. 2006. Global biodiversity conservation priorities. *Science* 313 (5783): 58–61.

Budd, G. E. 1998. Stem group arthropods from the Lower Cambrian Sirius Passet fauna of North Greenland. *Arthropod Relationships.* R. A. F. a. T. R. H. London: Systematics Association, Special Volume Series 55:125–38.

———. 2003. The Cambrian fossil record and the origin of the phyla. *Integrative and Comparative Biology* 43 (February): 157–15.

————, and S. Jensen. 2000. A critical reappraisal of the fossil record of the bilaterian phyla. *Biological Reviews of the Cambridge Philosophical Society* 75 (253–95).

Bush, A. M., and R. K. Bambach. 2004. Did alpha diversity increase during the Phanerozoic? Lifting the veils of taphonomic, latitudinal, and environmental biases. *Journal of Geology* 112:625–42.

Buss, L. W., and A. Seilacher. 1994. The phylum Vendobionta: A sister group of the Eumetazoa? *Paleobiology* 20:1–4.

Butterfield, N. 2003. Exceptional fossil preservation and the Cambrian explosion. *Integrative and Comparative Biology* 43:166–77.

Calcott, B. Forthcoming. Lineage explanations: Explaining evolutionary change without populations. *British Journal for the Philosophy of Science*.

Caldwell, C., and V. S. Johnston. 1991. Tracking a criminal suspect through "face-space" with a genetic algorithm. *Proceedings of the Fourth International Conference on Genetic Algorithm* (July): 416–21.

Callaway, R. 1997. Positive interactions in plant communities and the individualistic-continuum concept. *Oecologia* 112:143–49.

Callebaut, W., and D. Rasskin-Gutman. 2005. *Modularity: Understanding the development and evolution of natural complex systems*. Cambridge, MA: MIT Press.

Calvin, W. 2002. *A brain for all seasons: Human evolution and abrupt climate change*. Chicago: Univ. of Chicago Press.

Cardillo, M., G. Mace, et al. 2005. Multiple causes of high extinction risk in large mammal species. *Science* 309 (August 19): 1239–41.

————, G. Mace, et al. 2006. Latent extinction risk and the future battlegrounds of mammal conservation. *Proceedings of the National Academy of Science* 103:4157–61.

Caron, J. B., A. Scheltema, et al. 2006. A soft-bodied mollusc with radula from the Middle Cambrian Burgess Shale. *Nature* 442 (July 13): 159–63.

Carroll, R. L. 2002. Evolution of the capacity to evolve. *Journal of Evolutionary Biology* 15:911–21.

Carroll, S. B. 2005. *Endless forms most beautiful: The new science of evo devo and the making of the animal kingdom*. New York: W. W. Norton.

Caughley, G. 1995. Directions in conservation biology. *Journal of Animal Ecology* 63 (2): 215–44.

Cavalli-Sforza, L. L., P. Menozzi, and A. Piazza. 1994. *The history and geography of human genes*. Princeton, NJ: Princeton Univ. Press.

Chen, J.-Y., P. Oliveri, et al. 2000. Precambrian animal diversity: Putative phosphatized embryos from the Doushantuo Formation of China. *Proceedings of the National Academy of Science* 97 (9): 4457–62.

Claridge, M. F., H. A. Dawah, and M. R. Wilson. 1997. *Species: The units of biodiversity*. London: Chapman and Hall.

Clements, F. 1936. Nature and structure of climax. *Journal of Ecology* 24:252–84.

Cohan, F. M. 2002. What are bacterial species? *Annual Review of Microbiology* 56:457–87.

Condon, D., M. Zhu, et al. 2005. U-Pb ages from the Neoproterozoic Doushantuo Formation, China. *Science* 308:95–98.

Conway Morris, S. 1998. *The crucible of creation: The Burgess Shale and the rise of animals*. Oxford: Oxford Univ. Press.

———, J. S. Peel, et al. 1987. A Burgess Shale-like fauna from the Lower Cambrian of North Greenland. *Nature* 345:802–5.

Coope, G. R. 1994. The response of insect faunas to glacial-interglacial climatic fluctuations. *Philosophical Transactions of the Royal Society of London* B 344:19–26.

Cooper, G. 1993. The competition controversy in community ecology. *Biology and Philosophy* 8:359–84.

———. 2001. Must there be a balance of nature? *Biology and Philosophy* 16 (4): 481–506.

———. 2003. *The science of the struggle for existence*. Cambridge: Cambridge Univ. Press.

Cordell, G. A. 2000. Biodiversity and drug discovery—a symbiotic relationship. *Phytochemistry* 55:463–80.

Cracraft, J. 1983. Species concepts and speciation analysis. In *Current Ornithology*, 159–87. New York: Plenum Press.

Craft, A., and D. Simpson. 2001. The value of biodiversity in pharmaceutical research with differentiated products. *Environmental and Resource Economics* 18:1–17.

Cranston, P., P. Gullan, et al. 1991. Principles and practice of systematics. In *The insects of Australia*, ed. Division of Entomology, Commonwealth Scientific and Industrial Research Organisation, 109–24. Melbourne: Melbourne Univ. Press.

Cranston, P. S., and Hillman. 1992. Rapid assessment of biodiversity using biological diversity technicians. *Australian Biologist* 5 (3): 144–54.

Cummins, R. 1973. Functional analysis. *Journal of Philosophy* 72:741–64.

Dalton, R. 2004. Natural resources: Bioprospects less than golden. *Nature* 429:598–600.

Daugherty, C. H., A. Cree, et al. 1990. Neglected taxonomy and continuing extinctions of tuatara (Sphenodon). *Nature* 347:177–79.

Davidson, E., and D. H. Erwin. 2006. Gene regulatory networks and the evolution of animal body plans. *Science* 311 (February 10): 796–800.

Dawkins, R. 1976. *The selfish gene*. Oxford: Oxford Univ. Press.

———. 1986. *The Blind Watchmaker*. New York: W. W. Norton.

———. 1996. *Climbing Mount Improbable*. New York: W. W. Norton.

———. 2003. *A devil's chaplain: Reflections on hope, lies, science, and love*. Boston: Houghton Mifflin.

———. 2004. *The ancestor's tale: A pilgrimage to the dawn of life*. London: Weidenfeld and Nicholson.

———, and B. Goodwin. 1995. What is an organism? *Perspectives in Ethology* 2:47–60.

de Queiroz, K. 1998. The general lineage concept of species, species criteria, and the process of speciation: A conceptual unification and terminological recommendations. In *Endless Forms: Species and Speciation*, ed. D. J. Howard and S. H. Berlocher, 57–75. Oxford: Oxford Univ. Press.

———. 1999. The general lineage concept of species and the defining properties of the species category. In *Species: New Interdisciplinary Essays*, ed. R. A. Wilson, 49–89. Cambridge, MA: MIT Press.

———, and D. A. Good. 1997. Phenetic clustering in biology: A critique. *Quarterly Review of Biology* 72:3–30.

Delsuc, F., H. Brinkmann, and H. Philippe. 2005. Phylogenomics and the reconstruction of the tree of life. *Nature Reviews Genetics* 8 (May): 361–79.

Dennett, D. C. 1995. *Darwin's dangerous idea*. New York: Simon and Schuster.

Desmond, A. 1982. *Archetypes and ancestors: Palaeontology in Victorian London, 1850–1875*. London: Blond and Briggs.

Diamond, J. M. 1975. Assembly of species communities. In *Ecology and evolution of communities*, ed. M. L. Cody and J. M. Diamond, 342–444. Cambridge, MA: Harvard Univ. Press.

Díaz, S., J. Fargione, et al. 2006. Biodiversity loss threatens human well-being. *Public Library of Science (Biology)* 4 (8).

Douzery, E. J. P., E. A. Snell, et al. 2004. The timing of eukaryotic evolution: Does a relaxed molecular clock reconcile proteins and fossils? *Proceedings of the National Academy of Sciences* 101 (43): 15386–91.

Duffy, J. E. 2002. Biodiversity and ecosystem function: The consumer connection. *Oikos* 99 (2): 201–19.

Dupré, J. 1993. *The disorder of things: Metaphysical foundations of the disunity of science*. Cambridge, MA: Harvard Univ. Press.

———. 2001. In defence of classification. *Studies in the History and Philosophy of Biology and Biomedical Science* 32 (2): 203–19.

Eble, G. J. 2000. Contrasting evolutionary flexibility in sister groups: Disparity and diversity in Mesozoic atelostomate echinoids. *Paleobiology* 26:56–79.

Eermisse, D., and K. Peterson. 2004. The history of animals. In *Assembling the tree of life*, ed. J. Cracraft and M. Donoghue, 197–208. Oxford: Oxford Univ. Press.

Ehrenfeld, D. 1981. *The arrogance of humanism*. New York: Oxford Univ. Press.

Ehrlich, P. R., and A. H. Ehrlich. 1981. *Extinction: The causes and consequences of the disappearance of species*. New York: Random House.

———, and P. Raven. 1969. The differentiation of populations. *Science* 165:1228–32.

———, and B. Walker. 1998. Rivets and redundancy. *BioScience* 48:387.

Eldredge, N. 1995. *Reinventing Darwin: The great debate at the high table of evolutionary theory*. New York: Wiley.

———. 2003. The sloshing bucket: How the physical realm controls evolution. In *Evolutionary dynamics: Exploring the interplay of selection, accident, neutrality, and function*, ed. J. P. Crutchfield and P. Schuster, 3–32. Oxford: Oxford Univ. Press.

———, J. Thompson, et al. 2005. The dynamics of evolutionary stasis. *Paleobiology* 31 (2): 133–45.

Elton, C. S. 1927. *Animal ecology*. London: Sidgwick and Jackson.

Ereshefsky, M. 2001. *The poverty of the Linnaean hierarchy: A philosophical study of biological taxonomy*. Cambridge: Cambridge Univ. Press.

Ernest, S., K. Morgan, and J. H. Brown. 2001. Homeostasis and compensation: The role of species and resources in ecosystem stability. *Ecology* 82 (8): 2118–32.

Erwin, T. L. 1991. How many species are there? *Conservation Biology* 5:330–33.

Estes, S., and S. J. Arnold. 2007. Resolving the paradox of stasis: Models with stabilizing selection explain evolutionary divergence on all timescales. *American Naturalist* 169:227–44.

Faith, D. 1992. Conservation evaluation and phylogenetic diversity. *Biological Conservation* 61:1–10.

———. 1994. Phylogenetic pattern and the quantification of organismal biodiversity. *Philosophical Transactions of the Royal Society of London* B 345:45–58.

———. 2002. Quantifying biodiversity: A phylogenetic perspective. *Conservation Biology* 16 (1): 248–52.

———. 2003. Biodiversity. In *The Stanford Encyclopedia of Philosophy*, ed. E. N. Zalta, http://plato.stanford.edu/archives/sum2003/entries/biodiversity/.

Falkowski, P., and C. de Vargas. 2004. Shotgun sequencing in the sea: A blast from the past? *Science* 304 (April 2): 58–60.

Fierer, N., and R. Jackson. 2006. The diversity and biogeography of soil bacterial communities. *Proceedings of the National Academy of Science* 103 (3): 626–31.

Firn, R. D. 2003. Bioprospecting: Why is it so unrewarding? *Biodiversity and Conservation* 12 (2): 207–16.

———, and C. G. Jones. 2000. The evolution of secondary metabolism—a unifying model. *Molecular Microbiology* 37 (5): 989–94.

Fjeldså, J. 2000. The relevance of systematics in choosing priority areas for global conservation. *Environmental Conservation* 27:67–75.

Foote, M. 1996. Models of morphological diversification. In *Evolutionary Paleobiology*, ed. D. Jablonski, D. H. Erwin and J. H. Lipps, 62–86. Chicago: Univ. of Chicago Press.

———. 1997a. The evolution of morphological diversity. *Annual Review of Ecological Systematics* 28:129–52.

———. 1997b. Sampling, taxonomic description, and our evolving knowledge of morphological diversity. *Paleobiology* 23:181–206.

———. 2003. Origination and extinction through the Phanerozoic: A new approach. *Journal of Geology* 111:125–48.

———, and S. J. Gould. 1992. Cambrian and recent morphological disparity. *Science* 258:1816.

———, and S. Peters. 2001. Biodiversity in the Phanerozoic: A reinterpretation. *Paleobiology* 27 (4): 583–601.

Forest, F., R. Grenyer, et al. 2007. Preserving the evolutionary potential of floras in biodiversity hotspots. *Nature* 445 (February 15): 757–60.

Fortey, R. A., D. Briggs, et al. 1996. The Cambrian evolutionary "explosion":

Decoupling cladogenesis from morphological disparity. *Biological Journal of the Linnean Society* 57:13–33.

Friedberg, R. M. 1959. A learning machine. *IBM Journal of Research and Development* 3:183–91.

Futuyma, D. 1987. On the role of species in anagenesis. *American Naturalist* 130 (3): 465–73.

Gaston, K. 1996a. *Biodiversity: A biology of numbers and difference.* Oxford: Blackwell Science.

———. 1996b. Species richness: Measure and measurement. In *Biodiversity: A biology of numbers and difference,* 77–113. Oxford: Blackwell Science.

———, and T. Blackburn. 2000. *Pattern and process in macroecology.* Oxford: Blackwell.

———, and T. Blackburn. 1999. A critique for macroecology. *Okios* 84:353–68.

———, and J. I. Spicer. 2004. *Biodiversity: An introduction.* Malden, MA: Blackwell.

———, and P. H. Williams. 1993. Mapping the world's species—the higher taxon approach. *Biodiversity Letters* 1:2–8.

Gerhart, J. C., and M. W. Kirschner. 1997. *Cells, embryos, and evolution.* Malden, MA: Blackwell Science.

Gill, S., M. Pop, et al. 2006. Metagenomic analysis of the human distal gut microbiome. *Science* 312 (June 2): 1355–59.

Gilmour, J. S. L. 1940. Taxonomy and philosophy. In *The new systematics,* ed. J. Huxley, 401–74. Oxford: Oxford Univ. Press.

Gleason, H. A. 1926. The individualistic concept of plant associations. *Bulletin of the Torrey Botanical Club* 53:7–26.

Godfray, H. C. 2007. Linnaeus in the information age. *Nature* 446 (March 15): 259–60.

Godfrey-Smith, P. 1993. Functions: Consensus without unity. *Pacific Philosophical Quarterly* 74:196–208.

———. 1994. A modern history theory of functions. *Nous* 28 (344–62).

Goldenfeld, N., and C. Woese. 2007. Biology's next revolution. *Nature* 445 (January 25): 369.

Golley, F. B. 1993. *A history of the ecosystem concept in ecology: More than the sum of the parts.* New Haven, CT: Yale Univ. Press.

Gómez-Pompa, A. 2004. The role of biodiversity scientists in a troubled world. *BioScience* 54 (3): 217–25.

Goodfellow, M., G. P. Manfio, and J. Chun. 1997. Towards a practical species concept for cultivable bacteria. In *Species: The Units of Biodiversity,* ed. M. F. Claridge, H. A. Dawah, and M. R. Wilson, 25–59. London: Chapman and Hall.

Goodman, N. 1972. Seven strictures on similarity. In *Problems and projects,* 437–47. Indianapolis: Bobbs-Merrill.

Gould, J., and P. Marler. 1991. Learning by instinct. In *Behaviour and evolution of birds,* ed. D. Mock, 4–19. San Francisco, Freeman.

Gould, S. J. 1989. *Wonderful life.* New York: W. W. Norton.

———. 1991. The disparity of the Burgess Shale arthropod fauna and the limits of cladistic analysis: Why must we strive to quantify morphospace? *Paleobiology* 17:411–23.

———. 1993. How to analyze Burgess Shale disparity—A reply to Ridley. *Paleobiology* 19:522–23.

———. 2002. *Structure of evolutionary theory*. Cambridge, MA: Belknap Press of Harvard Univ. Press.

Gradstein, F., J. Ogg, et al. 2004. *A geologic time scale*. Cambridge: Cambridge Univ. Press.

Grantham, T. 2004. The role of fossils in phylogeny reconstruction: Why is it so difficult to integrate paleobiological and neontological evolutionary biology? *Biology and Philosophy* 19 (5): 687–720.

Greenslade, P. J. M., and P. Greenslade. 1984. Invertebrates in environmental assessment. *Environment and Planning Journal* 3:13–15.

Grelle, C. E. V. 2002. Is higher-taxon analysis a useful surrogate of species richness in studies of neotropical mammal diversity. *Biological Conservation* 108:101–6.

Grew, N. 1682. *The anatomy of plants*. London.

Griesemer, J. R. 1992. Niche: Historical perspectives. In *Keywords in evolutionary biology*, ed. E. F. Keller and E. A. Lloyd, 231–40. Cambridge, MA: Harvard Univ. Press.

Grotzinger, J., S. Bowring, et al. 1995. Biostratigraphic and geochronologic constraints on early animal evolution. *Science* 270:598–604.

Groves, C., D. Jensen, et al. 2002. Planning for biodiversity conservation: Putting conservation science into practice. *BioScience* 52 (5): 499–512.

Hammond, P. M. 1992. Species inventory. In *Global biodiversity: Status of the Earth's living resources*, ed. B. Groombridge, 17–39. London: Chapman and Hall.

———. 1994. Practical approaches to the estimation of the extent of biodiversity in speciose groups. *Philosophical Transactions of the Royal Society of London* B 345:119–36.

Hanski, I. 2004. Metapopulation theory, its use and misuse. *Basic and Applied Ecology* 5:225–29.

Harper, J. L., and D. L. Hawksworth, eds. 1995. *Biodiversity: Measurement and estimation*. London: Chapman and Hall.

Harvey, P., and M. Pagel. 1991. *The comparative method in evolutionary biology*. Oxford: Oxford Univ. Press.

Hedges, S. 2002. The origin and evolution of model organisms. *Nature Reviews Genetics* 3 (9): 838–84.

Helbig, A. J. 2005. Evolutionary genetics: A ring of species. *Heredity* 95:113–14.

Hendry, A. 2007. Evolutionary biology: The Elvis paradox. *Nature* 446 (March 8): 147–49.

Hennig, W. 1965. Phylogenetic systematics. *Annual Review of Entomology* 10:97–116.

Herbert, P., and T. R. Gregory. 2005. The promise of DNA barcoding for taxonomy. *Systematic Biology* 54 (5): 852–89.

Hey, J. 2001. *Genes, categories, and species: The evolutionary and cognitive cause of the species problem*. New York: Oxford Univ. Press.

Hoffmann, A. A., and J. Shirriffs. 2002. Geographic variation for wing shape in *Drosophila serrata*. *Evolution* 56:1068–73.

Holloway, J. D., and N. E. Stork. 1991. The dimensions of biodiversity: The use of invertebrates as indicators of human impact. In *The Biodiversity of Microorganisms and Invertebrates: Its Role in Sustainable Agriculture*, ed. D. I. Hawksworth, 37–62. Wallingford, England: C.A.B. International.

Hooper, D. J., F. S. Chapin, et al. 2005. Effects of biodiversity on ecosystem functioning: A consensus of current knowledge. *Ecological Monographs* 75 (1): 3–35.

Hou, X., R. Aldridge, et al. 2004. *The Cambrian fossils of Chengjiang, China: The flowering of early animal life.* Oxford: Blackwell.

———, and J. Bergstrom. 1995. Cambrian lobopodians—ancestors of extant onychophorans? *Zoological Journal of the Linnean Society* 114 (1): 3–19.

Hull, D. 1988. *Science as a process.* Chicago: Univ. of Chicago Press.

———. 1997. The ideal species concept—and why we can't get it. In *Species: The units of biodiversity*, ed. M. F. Claridge, H. A. Dawah, and M. R. Wilson, 357–80. London: Chapman and Hall.

———. 1999. On the plurality of species: Questioning the party line. In *Species: New interdisciplinary essays*, ed. R. A. Wilson, 23–48. Cambridge, MA: MIT Press.

———. 2006. Species. In *The philosophy of science: An encyclopedia*, ed. F. Pfeifer and S. Sarkar, 2:795–802. New York: Routledge.

Hulsey, D. C., and P. C. Wainwright. 2002. Projecting mechanics into morphospace: Disparity in the feeding system of labrid fishes. *Proceedings of the Royal Society of London* B 269:317–26.

Irwin, D. E., S. Bensch, et al. 2005. Speciation by distance in a ring species. *Science* 307 (5708): 414–16.

Jablonka, E., and M. J. Lamb. 2005. *Evolution in four dimensions: Genetic, epigenetic, behavioral, and symbolic variation in the history of life.* Cambridge, MA: MIT Press.

Jablonski, D. 2005. Mass extinctions and macroevolution. *Paleobiology* 31 (2): 192–210.

Jackson, F., and P. Pettit. 1992. In defense of explanatory ecumenism. *Economics and Philosophy* 8:1–21.

Jacob, F., and J. Monod. 1961. Genetic regulatory mechanisms in the synthesis of proteins. *Journal of Molecular Biology* 3:318–56.

Jax, K. 2006. Ecological units: Definitions and application. *Quarterly Review of Biology* 81 (3): 237–58.

Jenner, R., and M. Wills. 2007. The choice of model organisms in evo-devo. *Nature Reviews Genetics* 6:311–19.

Jermiin, L., L. Poladian, and M. Charleston. 2005. Is the "big bang" in animal evolution real? *Science* 310 (December): 1910–11.

John, C. M., and C. A. Maggs. 1997. Species problems in eukaryotic algae: A modern perspective. In *Species: The units of biodiversity*, ed. M. F. Claridge, H. A. Dawah, and M. R. Wilson, 83–108. London: Chapman and Hall.

Jones, C., J. Lawton, and M. Shachak. 1997. Positive and negative effects of organisms as physical ecosystems engineers. *Ecology* 78:1946–57.

Jordán, F., and I. Scheuring. 2002. Searching for keystones in ecological networks. *Oikos* 99 (3): 607–12.

Kauffman, S. 1993. *The origins of order: Self-organisation and selection in evolution.* New York: Oxford Univ. Press.

———. 1995. *At home in the universe.* New York: Oxford Univ. Press.

Kendall, D. 1977. The diffusion of shape. *Advances in Applied Probability* 9:428–30.

Kiessling, W. 2005. Long-term relationships between ecological stability and biodiversity in Phanerozoic reefs. *Nature* 433 (January 27): 410–13.

Kingsland, S. 1985. *Modeling nature: Episodes in the history of population ecology.* Chicago: Univ. of Chicago Press.

Kinnear, J. E., N. R. Sumner, and M. L. Onus. 2002. The red fox in Australia—an exotic predator turned biocontrol agent. *Biological Conservation* 108 (3): 335–59.

Kinzig, A. P., S. W. Pacala, and D. Tilman, eds. 2001. *The functional consequences of biodiversity.* Princeton, NJ: Princeton Univ. Press.

Kirschner, M., and J. C. Gerhart. 1998. Evolvability. *Proceedings of the National Academy of Science* 95:8420–827.

———, J. Gerhart, and J. Norton. 2005. *The plausibility of life: Resolving Darwin's dilemma.* New Haven, CT: Yale Univ. Press.

Kirschvink, J., and T. Raub. 2003. A methane fuse for the Cambrian explosion: Carbon cycles and true polar wander. *Comptes. Rendus Geosciences* 335:65–78.

Kitcher, P. 1984a. Against the monism of the moment: A reply to Elliott Sober. *Philosophy of Science* 51:616–30.

———. 1984b. Species. *Philosophy of Science* 51:308–33.

Knoll, A. H. 2003. *Life on a young planet: The first three billion years of evolution on earth.* Princeton, NJ: Princeton Univ. Press.

Landing, E., S. Bowring, et al. 1998. Duration of the early Cambrian: U-Ph ages of volcanic ashes from Avalon and Gondwana. *Canadian Journal of Earth Sciences* 35:329–38.

Lauder, G. V. 1995. On the inference of function from structure. In *Functional morphology in vertebrate paleontology,* ed. J. Thomason, 1–18. Cambridge: Cambridge Univ. Press.

Lee, M. S. Y. 1992. Cambrian and recent morphological disparity. *Science* 258:1816–17.

Lehman, C., and D. Tilman. 2000. Biodiversity, stability, and productivity in competitive communities. *American Naturalist* 156 (5): 534–52.

Leopold, A. 1949. *A Sand County almanac.* New York: Oxford Univ. Press.

Lévêque, C., and J.-C. Mounolou. 2003. *Biodiversity.* Chichester, England: John Wiley.

Levins, R., and R. Lewontin, eds. 1985. *The dialectical biologist.* Cambridge, MA: Harvard Univ. Press.

Lewontin, R. 1985. Adaptation. In *The dialectical biologist,* ed. R. Levins and R. Lewontin, 65–84. Cambridge, MA: Harvard Univ. Press.

Liebers, D., P. de Knijff, and A. J. Helbig. 2004. The herring gull complex is not a ring species. *Proceedings of the Royal Society of London* B 271:893–901.

Light, A., and H. Rolston. 2003. *Environmental ethics: An anthology*. Oxford: Blackwell.

Littlewood, D., M. Telford, and R. Bray. 2004. Protostomes and Platyhelminthes: The worm's turn. In *Assembling the tree of life*, ed. J. Cracraft, M. Donoghue, and R. Bray, 209–34. Oxford: Oxford Univ. Press.

Lofgren, A. S., R. E. Plotnick, and P. Wagner. 2003. Morphological diversity of carboniferous arthropods and insights on disparity patterns through the Phanerozoic. *Paleobiology* 29 (3): 349–68.

Lomborg, B. 2001. *The skeptical environmentalist: Measuring the real state of the world*. Cambridge: Cambridge Univ. Press.

Loreau, M., S. Naeem, et al. 2001. Biodiversity and ecosystem functioning: Current knowledge and future challenges. *Science* 294 (October 26): 804–8.

Lyons, K. G., C. A. Brigham, et al. 2005. Rare species and ecosystem functioning. *Conservation Biology* 19 (4): 1019–24.

Mace G. M., J. L. Gittleman, and A. Purvis. 2003. Preserving the tree of life. *Science* 300:1707–9.

Macilwain, C. 1998. When rhetoric hits reality in debate on bioprospecting. *Nature* 392:535–40.

Maclaurin, J. 2003. The good, the bad, and the impossible. *Biology and Philosophy* 18 (3): 463–76.

Majer, J. D. 1983. Ant: Bio-indicators of minesite rehabilitation, land-use, and land conservation. *Environmental Management* 7:375–83.

Mallet, J. 1996. The genetics of biological diversity: From varieties to species. In *Biodiversity: A biology of numbers and difference*, ed. K. Gaston, 1–9. Cambridge, MA: Blackwell Science.

———. 2007. Hybrid speciation. *Nature* 446 (March 15): 279–83.

Maloof, A., D. Schrag, et al. 2005. An expanded record of early Cambrian carbon cycling from the Anti-Atlas Margin, Morocco. *Canadian Journal of Earth Sciences* 42 (2): 195–216.

Margules, C. R., and R. L. Pressey. 2000. Systematic conservation planning. *Nature* 405 (May 11): 243–53.

Marris, E. 2007. The species and the specious. *Nature* 446 (March 15): 250–53.

Marshall, C. 2006. Explaining the Cambrian "explosion" of animals. *Annual Review of Earth and Planetary Sciences* 34:355–84.

Marshall, L. G. 1981. The great American interchange: An invasion induced crisis for South American mammals. In *Biotic crises in ecological and evolutionary time*, ed. M. H. Nitecki, 133–229. New York: Academic Press.

May, R. M. 1973. *Stability and complexity in modal ecosystems*. Princeton, NJ: Princeton Univ. Press.

———. 1992. How many species inhabit the Earth? *Scientific American* 264 (4): 18–24.

———. 1995. Conceptual aspects of the quantification of the extent of biological diversity. In *Biodiversity: Measurement and estimation*, ed. D. L. Hawksworth, 13–21. London: Chapman and Hall.

Mayr, E. 1942. *Systematics and the origin of species*. New York: Columbia Univ. Press.

———. 1969. *Principles of systematic zoology.* New York: McGraw-Hill.

McCann, K. 2007. Protecting biostructure. *Nature* 446 (March 1): 29.

McGhee, G. R. 1999. *Theoretical morphology: The concept and its applications.* New York: Columbia Univ. Press.

———. 2006. *The geometry of evolution: Adaptive landscapes and theoretical morphospaces.* Cambridge: Cambridge Univ. Press.

McGinnis, W., M. S. Levine, et al. 1984. A conserved DNA sequence in homeotic genes of the *Drosophila* antennapedia and bithoras complexes. *Nature* 308:428–33.

McGowan, A. J. 2004. Ammonoid taxonomic and morphologic recovery patterns after the Permian-Triassic. *Geology* 32:665–68.

McMenamin, M., and D. McMenamin. 1990. *The emergence of animals: The Cambrian breakthrough.* New York: Columbia Univ. Press.

McShea, D. W. 1992. A metric for the study of evolutionary trends in the complexity of serial structures. *Biological Journal of the Linnean Society of London* 43:39–55.

———. 1993. Arguments, tests, and the Burgess Shale—A commentary on the debate. *Paleobiology* 19 (4): 399–402.

———. 1996. Metazoan complexity and evolution: Is there a trend? *Evolution* 50 (2): 477–92.

Meyers, N. 1979. *The sinking ark: A new look at the problem of disappearing species.* New York: Pergamon Press.

Millikan, R. 1989. In defense of proper functions. *Philosophy of Science* 56:288–302.

Mishler, B. D. 1999. Getting rid of species? In *Species: New interdisciplinary essays,* ed. R. A. Wilson, 307–15. Cambridge, MA: MIT Press.

———, and M. J. Donoghue. 1982. Species concepts: A case for pluralism. *Systematic Zoology* 31:491–503.

Montañez, I., D. Osleger, et al. 2000. Evolution of Sr and C isotope composition of Cambrian oceans. *GSA Today* 10 (5): 1–7.

Montoya, J., M. Rodriguez, and B. Hawkins. 2003. Food web complexity and higher-level ecosystem services. *Ecology Letters* 6:587–93.

Mooers, A. 2007. The diversity of biodiversity. *Nature* 445 (February 15): 717–18.

Moritz, C. 1995. Uses of molecular phylogenies for conservation. *Philosophical Transactions of the Royal Society of London* B 349:113–18.

Morjan, C., and L. Rieseberg. 2004. How species evolve collectively: Implications of gene flow and selection for the spread of advantageous alleles. *Molecular Ecology* 13:1341–56.

Murdoch, W. W. 1994. Population regulation in theory and practice. *Ecology* 75 (2): 271–87.

Naeem, S. 1998. Species redundancy and ecosystem reliability. *Conservation Biology* 12 (1): 39–45.

———, and A. C. Baker. 2005. Paradise sustained. *Nature* 433 (January 27): 370–71.

Naido, R., and T. Ricketts. 2006. Mapping the economic costs and benefits of conservation. *Public Library of Science (Biology)* 4 (11): e360.

Naiman, R. J., J. M. Milillo, and J. E. Hobbie. 1986. Ecosystem alteration of boreal forest streams by beaver (*Castor canadensis*). *Ecology* 67:1254–369.

Narbonne, G. 1998. The Ediacaran biota: A terminal Proterozoic experiment in the evolution of life. *GSA Today* 8 (2): 1–7.

Nash, R. 1990. *The rights of nature: A history of environmental ethics.* Leichhardt, Australia: Primavera Press.

Neige, P. 2003. Spatial patterns of disparity and diversity of the recent cuttlefishes (Cephalopoda) across the Old World. *Journal of Biogeography* 30:1125–37.

Newman, D. J., G. M. Cragg, and K. Snader. 2003. Natural products as sources of new drugs over the period 1981–2002. *Journal of Natural Products* 66 (7): 1022–37.

Niklas, K. J. 1994. Morphological evolution through complex domains of fitness. *Proceedings of the National Academy of Sciences* 91 (July 1994): 6772–79.

———. 2004. Computer models of early land plant evolution. *Annual Review of Earth and Planetary Sciences* 32:47–66.

Nilsson, D.-E., and S. Pelger. 1994. A pessimistic estimate of the time required for an eye to evolve. *Proceedings of the Royal Society of London,* B 256:53–58.

Norton, B. 1987. *Why preserve natural variety?* Princeton, NJ: Princeton Univ. Press.

———. 2003. Environmental ethics and weak anthropocentrism. In *Environmental ethics: An anthology,* ed. A. Light and H. Rolston, 163–74. Oxford: Blackwell.

Odenbaugh, J. 2006. Ecology. In *The Philosophy of Science: An Encyclopedia,* ed. S. Sarkar and F. Pfeifer, 215–24. New York: Routledge.

Odling-Smee, F. J., K. N. Laland, and M. Feldman. 2003. *Niche construction: The neglected process in evolution.* Princeton, NJ: Princeton Univ. Press.

Ohno, S. 1996. The notion of the Cambrian pananimalia genome. *Proceedings of the National Academy of Science* 93:8475–78.

O'Malley, M., and J. Dupré. 2007. Size doesn't matter: Towards a more inclusive philosophy of biology. *Biology and Philosophy* 22 (2): 155–91.

Orr, H. A. 2005. The genetic theory of adaptation: A brief history. *Nature Reviews Genetics* 6:119–27.

Owens, I. P. F., and P. M. Bennett. 2000. Quantifying biodiversity: A phenotypic perspective. *Conservation Biology* 14:1014–22.

Owen-Smith, N. 1988. *Megaherbivores: The influence of very large body size on ecology.* Cambridge: Cambridge Univ. Press.

Oyama, S., P. Griffiths, and R. Gray, eds. 2001. *Cycles of contingency: Developmental systems and evolution.* Cambridge, MA: MIT Press.

Paine, R. T. 1966. Food web complexity and species diversity. *American Naturalist* 100:65–75.

Parker, V. T. 2004. The community of an individual: Implications for the community concept. *Oikos* 104:27–34.

Paterson, H. E. H. 1985. The recognition concept of species. In *Species and speciation,* ed. E. Vrba, 21–29. Pretoria: Transvaal Museum.

Pearce, D. W., and S. Puroshothaman. 1995. The economic value of plant-based pharmaceuticals. In *Intellectual property rights and biodiversity conservation,* ed. T. Swanson, 127–38. Cambridge: Cambridge Univ. Press.

Pearson, D. L. 1995. Selecting indicator taxa for the quantitative assessment of biodiversity. In *Biodiversity: Measurement and estimation*, 75–79. London: Chapman and Hall.

Peterson, K., J. Lyons, et al. 2004. Estimating metazoan divergence times with a molecular clock. *Proceedings of the National Academy of Science* 101 (17): 6536–41.

Pimm, S. 1991. *The balance of nature*. Chicago: Univ. of Chicago Press.

Plotnick, R. E., and T. K. Baumiller. 2000. Invention by evolution: Functional analysis in paleobiology. *Paleobiology* 26 (4 Supplement): 305–23.

Pomeroy, L. 2001. Caught in the food web: Complexity made simple? *Scientia Marina* 65 (2 Supplement): 31–40.

Posadas, P., D. R. Miranda-Esquivel, and J. V. Crisci. 2004. On words, tests, and applications: Reply to Faith et al. *Conservation Biology* 18 (1): 262–66.

Power, M. E., D. Tilman, et al. 1996. Challenges in the quest for keystones. *BioScience* 46 (8): 609–20.

Prance, G. T. 1994. A comparison of the efficacy of higher taxa and species numbers in the assessment of biodiversity in the neotropics. *Philosophical Transactions of the Royal Society of London* 345:89–99.

Prusinkiewicz, P., and A. Lindenmayer. 1990. *The algorithmic beauty of plants*. New York: Springer-Verlag.

Puppe, C., and K. Nehring. 2002. A theory of diversity. *Econometrica* 70:1155–98.

———. 2004. Modelling phylogenetic diversity. *Resource and Energy Economics* 26:205–35.

Purvis, A., and A. Hector. 2000. Getting the measure of biodiversity. *Nature* 405:212–19.

Quicke, D. 1993. *Principles and techniques of contemporary taxonomy*. London: Blackie Academic and Professional.

Raff, R. A. 1996. *The shape of life: Genes, development, and the evolution of animal form*. Chicago: Univ. of Chicago Press.

Raup, D. 1966. Geometric analysis of shell coiling: General problems. *Journal of Paleontology* 40:1178–90.

———. 1979. Size of the Permo-Triassic bottleneck and its evolutionary implications. *Science* 206:217–18.

———. 1991. *Extinction: Bad genes or bad luck?* Oxford: Oxford Univ. Press.

Reif, W. E. 1980. A model of morphogenetic processes in the dermal skeleton of elasmobranchs. *Neues Jahrbuch für Geologie und Paläontologie, Abhandlungen* 159:339–59.

Reisman, K., and P. Forber. 2005. Manipulation and the causes of evolution. *Philosophy of Science* 72:1113–23.

Ricklefs, R. E. 2004. A comprehensive framework for global patterns in biodiversity. *Ecology Letters* 7 (1): 1–15.

———, and D. B. Miles. 1994. Ecological and evolutionary inferences from morphology: An ecological perspective. In *Ecological morphology*, ed. P. C. Wainwright and S. M. Reilly, 13–41. Chicago: Univ. of Chicago Press.

———, and D. Schluter. 1993. Species diversity: Regional and historical influences. In *Species diversity in ecological communities*, ed. R. E. Ricklefs and D. Schluter, 350–63. Chicago: Univ. of Chicago Press.

Ridley, M. 1986. *Evolution and classification: The reformation of cladism.* London: Longman.

———. 1989. The cladistic solution to the species problem. *Biology and Philosophy* 4:1–16.

———. 1990. Dreadful beasts. *London Review of Books:* 11–12.

———. 1993. Analysis of the Burgess Shale. *Paleobiology* 19:519–21.

Rokas, A., D. Krüger, and S. Carroll. 2005. Animal evolution and the molecular signature of radiations compressed in time. *Science* 310 (December 23): 1933–38.

Rolston, H. 2001. Biodiversity. In *A companion to environmental philosophy*, ed. D. Jamieson, 403–15. Oxford: Blackwell.

Rosenberg, A. 2006. *Darwinian reductionism; Or, How to stop worrying and love molecular biology.* Chicago: Univ. of Chicago Press.

Roughgarden, J. 1995. *Anolis lizards of the Caribbean: Ecology, evolution, and plate tectonics.* Oxford: Oxford Univ. Press.

———, and S. Pacala. 1989. Taxon cycle among anolis lizard populations: Review of evidence. In *Speciation and its consequences*, ed. D. Otte and J. A. Endler, 403–32. Sunderland, MA: Sinauer.

Roy, K., D. P. Balch, and M. Hellberg. 2001. Spatial patterns of morphological diversity across the Indo-Pacific: Analyses using strombid gastropods. *Proceedings of the Royal Society of London* B 268:2503–8.

———, and M. Foote. 1997. Morphological approaches to measuring biodiversity. *Trends in Ecology and Evolution* 12 (7): 277–81.

———, D. Jablonski, and J. Valentine. 2004. Beyond species richness: Biogeographic patterns and biodiversity dynamics using other metrics of diversity. In *Frontiers of biogeography*, ed. M. V. Lomolino and L. R. Heaney, 151–70. Sunderland, MA: Sinauer.

Rupke, N. 1994. *Richard Owen: Victorian naturalist.* New Haven, CT: Yale Univ. Press.

Rutherford, S. 2000. From genotype to phenotype: Buffering mechanisms and the storage of genetic information. *BioEssays* 22 (12): 1095–105.

Ryan Gregory, T. 2001. Coincidence, coevolution, or causation? DNA content, cell size, and the C-value enigma. *Biological Reviews of the Cambridge Philosophical Society* 76:65–101.

Sanderson, M. J., and L. Hufford, eds. 1996. *Homoplasy: The recurrence of similarity in evolution.* San Diego: Academic Press.

Sandin, P. 2005. Dimensions of the precautionary principle. *Human and Ecological Risk Assessment* 5 (5): 889–907.

———, M. Peterson, et al. 2002. Five charges against the precautionary principle. *Journal of Risk Research* 5 (4): 287–99.

Sarkar, S. 2002. Defining biodiversity; assessing biodiversity. *Monist* 85 (1): 131–55.

———. 2005. *Biodiversity and environmental philosophy*. Cambridge: Cambridge Univ. Press.

Savolainen, V., R. Cowan, et al. 2005. Towards writing the encyclopedia of life: An introduction to DNA barcoding. *Philosophical Transactions of the Royal Society of London B* 360 (1462): 1805–11.

Schidlowski, M. 1988. A 3,800 million year isotopic record of life from carbon in sedimentary rocks. *Nature* 333:313–18.

Schlichting, C., and M. Pigliucci. 1998. *Phenotypic evolution: A reaction norm perspective*. Sunderland, MA: Sinauer.

Schliewen, U. K., and B. Klee. 2004. Reticulate sympatric speciation in Cameroonian crater lake cichlids. *Frontiers in Zoology* 1:56–67.

Schlosser, G., and G. P. Wagner. 2004. *Modularity in development and evolution*. Chicago: Univ. of Chicago Press.

Schluter, D. 2000. *The ecology of adaptive radiation*. Oxford: Oxford Univ. Press.

Seehausen, O. 2004. Hybridization and adaptive radiation. *Trends in Ecology and Evolution* 19 (4): 198–207.

Seilacher, A., and L. Buss. 1994. The phylum Vendobionta: A sister group of the Eumetazoa? *Paleobiology* 20:1–4.

Sepkoski, J. J. 1997. Biodiversity: Past, present, and future. *Journal of Paleontology* 71 (4): 533–39.

———. 2002. *A compendium of fossil marine animal genera*. Bulletins of American Paleontology. Paleontological Research Institution, Ithaca, NY, no. 363.

Sheail, J. 1976. Nature in trust: The history of nature conservation in Britain. Glasgow: Blackie.

Simpson, R. D., and R. Sedjo. 2004. Golden rule of economics yet to strike prospectors. *Nature* 430:723.

Simpson, G. G. 1944. *Tempo and mode in evolution*. New York: Columbia Univ. Press.

———. 1953. *The major features of evolution*. New York: Columbia Univ. Press.

Smith, A., and K. Peterson. 2002. Dating the time of origin of major clades: Molecular clocks and the fossil record. *Annual Review of Earth and Planet Science* 30:65–88.

Smith, V. 2005. DNA barcoding: Perspectives from a "Partnerships for Enhancing Expertise in Taxonomy" (PEET) debate. *Systematic Biology* 54 (5): 841–44.

Sober, E. 1984. *The nature of selection: Evolutionary theory in philosophical focus*. Cambridge, MA: MIT Press.

———. 1986. Philosophical problems for environmentalism. In *The preservation of species*, ed. B. Norton, 173–94. Princeton, NJ: Princeton Univ. Press.

Sokal, R. 1985. The continuing search for order. *American Naturalist* 126:729–47.

———, and P. H. Sneath. 1963. *The principles of numerical taxonomy*. San Francisco: W. H. Freeman.

Soulé, M. 1985. What is conservation biology? *BioScience* 35:727–34.

———, J. A. Estes, et al. 2003. Ecological effectiveness: Conservation goals for interactive species. *Conservation Biology* 17 (5): 1238–50.

Stanley, S. 1990. Delayed recovery and the spacing of major extinctions: Paleobiology. *Paleobiology* 16:401–14.

Steadman, D. A., and P. S. Martin. 2003. The Late Quaternary extinction and future resurrection of birds on Pacific Islands. *Earth-Science Reviews* 61:133–47.

Sterelny, K. 1996. Explanatory pluralism in evolutionary biology. *Biology and Philosophy* 11:193–214.

———. 2001a. The reality of ecology assemblages: A palaeo-ecological puzzle. *Biology and Philosophy* 16 (4): 437–61.

———. 2001b. Niche construction, developmental systems, and the extended replicator. In *Cycles of contingency*, ed. S. Oyama, R. D. Gray, and P. Griffiths, 33–49. Cambridge, MA: MIT Press.

———. 2004. Symbiosis, evolvability, and modularity. In *Modularity in development and evolution*, ed. G. Schlosser and G. Wagner, 490–516. Chicago: Univ. of Chicago Press.

———. 2007. Macroevolution, minimalism, and the radiation of the animals. In *Cambridge companion to the philosophy of biology*, ed. D. Hull and M. Ruse, 182–210. Cambridge: Cambridge Univ. Press.

———, and P. E. Griffiths. 1999. *Sex and death: An introduction to philosophy of biology*. Chicago: Univ. of Chicago Press.

Stokes, D. L. 2007. Things we like: Human preferences among similar organisms and implications for conservation. *Human Ecology* 35:361–69.

Sunstein, C. R. 2005. *Laws of fear: Beyond the precautionary principle*. Cambridge: Cambridge Univ. Press.

Templeton, A. 1989. The meaning of species and speciation: A genetic perspective. In *Speciation and its consequences*, ed. D. Otte and J. Endler, 3–27. Sunderland, MA: Sinauer.

———. 1998. Species and speciation: Geography, population structure, ecology, and gene trees. in *Endless forms: Species and speciation*, ed. D. J. Howard and S. H. Berlocher, 32–43. Oxford: Oxford Univ. Press.

Thebault, E., and M. Loreau. 2003. Food-web constraints on biodiversity-ecosystem functioning relationships. *Proceedings of the National Academy of Science* 100 (25): 14949–54.

Thomas, R. D. K., and W. E. Reif. 1993. The skeleton space: A finite set of organic designs. *Evolution* 47:341–60.

———, R. H. Shearman, and G. W. Stewart. 2000. Evolutionary exploitation of design options by the first animals with hard skeletons. *Science* 288:1239–42.

Thompson, D. W. 1917. *On growth and form*. Cambridge: Cambridge Univ. Press.

———. 1942. *On Growth and Form*, new ed. Cambridge: Cambridge Univ. Press; New York: Macmillan.

Thompson, J. N. 1999. The raw material for coevolution. *Oikos* 84:5–16.

———. 2005. *Geographic mosaic of coevolution*. Chicago: Univ. of Chicago Press.

Tilman, D. 1996. Biodiversity: Population versus ecosystem stability. *Ecology* 77 (2): 350–63.

———. 1999. The ecological consequences of changes in biodiversity: A search for general principles. *Ecology* 80 (5): 1455–74.

———, C. L. Lehman, and C. E. Bristow. 1998. Diversity-stability relationships: Statistical inevitability or ecological consequence? *American Naturalist* 151:277.

———, C. L. Lehman, and K. T. Thomson. 1997. Plant diversity and ecosystem productivity: Theoretical considerations. *Proceedings of the National Academy of Science* 94:1857–61.

———, S. Polasky, and C. Lehman. 2005. Diversity, productivity, and temporal stability in the economies of humans and nature. *Journal of Environmental Economics and Management* 49:405–26.

———, P. Reich, and J. M. H. Knops. 2006. Biodiversity and ecosystem stability in a decade-long grassland experiment. *Nature* 441:629–32.

Tomasello, M. 1999. *The cultural origins of human cognition.* Cambridge, MA: Harvard Univ. Press.

Tomlinson, A. 1973. Meteorological aspects of trans-Tasman insect dispersal. *New Zealand Entomologist* 5:253–68.

Trewick, S. A. 1997. Flightlessness and phylogeny amongst endemic rails (Aves: Rallidae) of the New Zealand region. *Philosophical Transactions of the Royal Society of London* B 352:429–46.

Valentine, J. 2004. *On the origin of phyla.* Chicago: Univ. of Chicago Press.

———, A. Collins, and C. Porter Meyer. 1994. Morphological complexity increase in metazoans. *Paleobiology* 20:131–42.

———, and D. Jablonski. 1993. Fossil communities: Compositional variation on many time scales. In *Species diversity in ecological communities,* ed. R. E. Ricklets and D. Schluter, 341–49. Chicago: Univ. of Chicago Press.

———, K. Roy, and D. Jablonski. 2002. Carnivore/non-carnivore ratios in northeastern Pacific marine gastropods. *Marine Ecology Progress Series* 228:153–63.

van Kooten, G. C., and E. H. Bulte. 2000. *The economics of nature: Managing biological assets.* Oxford: Blackwell.

van Valen, L. 1976. Ecological species, multispecies, and oaks. *Taxon* 25:233–39.

Vane-Wright, R. I., C. J. Humphries, and P. H. Williams. 1991. What to protect— systematics and the agony of choice. *Biological Conservation* 55:235–54.

Vermeij, G. J. 1999. Inequality and the directionality of history. *American Naturalist* 153 (3): 243–53.

Vizcaíno, S. F., and G. D. Iuliis. 2003. Evidence for advanced carnivory in fossil armadillos (Mammalia: Xenarthra: Dasypodidae). *Paleobiology* 29:123–38.

Vrba, E. 1993. Turnover-pulses, the red queen and related topics. *American Journal of Science* 293 (A): 418–52.

———. 1995. Species as habitat-specific complex systems. In *Speciation and the recognition concept: Theory and application,* ed. D. M. Lambert and H. Spencer, 3–44. Baltimore: Johns Hopkins Univ. Press.

Waddington, C. H. 1942. The canalization of development and the inheritance of acquired characters. *Nature* 150:563–65.

Wagner, G., and L. Altenberg. 1996. Complex adaptations and the evolution of evolvability. *Evolution* 50 (3): 967–76.

———, and J. G. Mezey. 2004. The role of genetic architecture constraints in the origin of variational modularity. In *Modularity in development and evolution*, ed. G. Schlosser and G. P. Wagner, 338–58. Chicago: Univ. of Chicago Press.

———, J. G. Mezey, and R. Calabretta. 2005. Natural selection and the origin of modules. In *Modularity: Understanding the development and evolution of natural complex systems*, ed. W. Callebaut and D. Rasskin-Gutman, 33–50. Cambridge, MA: MIT Press.

Wagner, P. 1995a. Diversity patterns amongst early gastropods: Contrasting taxonomic and phylogenetic descriptions. *Paleobiology* 21:410–39.

———. 1995b. Testing evolutionary constraint hypotheses with early Paleozoic gastropods. *Paleobiology* 21:248–72.

Walker, B. H. 1992. Biodiversity and ecological redundancy. *Conservation Biology* 6:18–23.

———. 1995. Conserving biological diversity through ecosystem resilience. *Conservation Biology* 9:1–7.

Ward, P. 1980. Comparative shell shape distribution in Jurassic-Cretaceous ammonites and Jurassic-Tertiary nautilids. *Paleobiology* 6 (1): 32–43.

Ward, S., and I. Thornton. 2000. Chance and determinism in the development of isolated communities. *Global Ecology and Biogeography* 9 (1): 7–18.

Wardle, D. A. 1999. Is "sampling effect" a problem for experiments investigating biodiversity-ecosystem function relationships? *Oikos* 87 (2): 403–7.

———, M. A. Huston, et al. 2000. Biodiversity and ecosystem function: An issue in ecology. *Ecological Society of America Bulletin* 81 (3): 232–35.

Weir, J. T., and D. Schluter. 2007. The latitudinal gradient in recent speciation and extinction rates of birds and mammals. *Science* 315:1574–76.

Wenz, P. 2001. *Environmental ethics today*. Oxford: Oxford Univ. Press.

West-Eberhard, M. J. 2003. *Developmental plasticity and evolution*. Oxford: Oxford Univ. Press.

Wheeler, Q., and R. Meier. 2000. *Species concepts and phylogenetic theory*. New York: Columbia Univ. Press.

Whitfield, J. 2004. Geology: "Time Lords." *Nature* 429 (13 May): 124–25.

Whittaker, R. H. 1975. *Communities and ecosystems*. New York: Macmillan.

Wiley, E. O. 1978. The evolutionary species concept reconsidered. *Systematic Zoology* 27:17–26.

Wilkins, J. 2007. The dimensions, modes, and definitions of species and speciation. *Biology and Philosophy* 22 (2): 247–66.

Will, K., B. D. Mishler, and Q. D. Wheeler. 2005. The perils of DNA barcoding and the need for integrative taxonomy. *Systematic Biology* 54 (5): 844–51.

Williams, G. C. 1992. *Natural selection: Domains, levels, and challenges*. New York: Oxford Univ. Press.

———. 1997. *The pony fish's glow: And other clues to plan and purpose in nature*. New York: Basic Books.

Williams, P., K. Gaston, and C. Humphries. 1994. Do conservationists and molecular biologists value differences between organisms in the same way? *Biodiversity Letters* 2:67–78.

———, and C. Humphries. 1996. Comparing character diversity among biotas. In *Biodiversity: A biology of numbers and difference*, ed. K. Gaston, 54–76. Oxford: Blackwell Science.

Wilson, D. S. 1997. Biological communities as functionally organized units. *Ecology* 78:2018–224.

Wilson, E. O. 1992. *The diversity of life*. Cambridge, MA: Belknap Press of Harvard Univ. Press.

———, ed. 1988. *Biodiversity*. Washington, DC: National Academy Press.

Wilson, R. A. 1999. *Species: New interdisciplinary essays*. Cambridge, MA: MIT Press.

Wimsatt, W. C. 2001. Generative entrenchment and the developmental systems Approach to evolutionary processes. In *Cycles of contingency: Developmental systems and evolution*, ed. R. Gray, P. Griffiths, and S. Oyama, 219–38. Cambridge, MA: MIT Press.

———. 2007. *Re-engineering philosophy for limited beings*. Cambridge, MA: Harvard Univ. Press.

———, and J. C. Schank. 1988. Two constraints on the evolution of complex adaptations and the means of their avoidance. In *Evolutionary progress*, ed. M. H. Nitecki, 231–75. Chicago: Univ. of Chicago Press.

Woodward, J. 2002. *A theory of explanation: Causation, invariance, and intervention*. Oxford: Oxford Univ. Press.

———. 2003. *Making things happen: A theory of causal explanation*. Oxford: Oxford Univ. Press.

Worm, B., and J. E. Duffy. 2003. Biodiversity, productivity and stability in real food webs. *Trends in Ecology and Evolution* 18 (2): 628–32.

Worster, D. 1994. *Nature's economy: A history of ecological ideas*. Cambridge: Cambridge Univ. Press.

Wray, G., J. Levinton, and L. Shapiro. 1996. Molecular evidence for deep Precambrian divergences amongst metazoan phyla. *Science* 274:568–73.

Wright, L. 1973. Functions. *Philosophical Review* 82:139–68.

Yang, A. 2001. Modularity, evolvability, and adaptive radiations: A comparison of the hemi- and holometabolous insects. *Evolution and Development* 3 (2): 59–72.

Zelditch, M. L., D. L. Swiderski, et al. 2004. *Geometric morphometrics for biologists: A primer*. New York: Elsevier.

Zimmer, C. 2007. Jurassic genomes. *Science* 315 (5817): 1358–59.

Index

Wagner, Peter, 101–2, 183n9
Walcott, Charles, 46
Walker, Brian, 168
Wentworth Thompson, D'Arcy, 64, 65, 79
West-Eberhard, Mary Jane, 91, 92, 93, 94, 157, 180n4
What to Protect—Systematics and the Agony of Choice (Vane-Wright), 161–62
white gum (*Eucalyptus rossii*), 120
white-tailed deer (*Odocoileus virginianus*), 169

white-winged chough (*Corcorax melanorhamphos*), 124
Whittington, Harry, 46, 54
Wilkins, John, 34
Wilson, David, 122
Wilson, Edward O., 1–2
Wimsatt, William, 55, 95, 96, 97, 102–3
Wonderful Life (Gould), 43, 44, 49, 53
Wright, Sewall, 34

Yang, Andrew, 103